CONDUCTIVE POLYMERS

CONDUCTIVE POLYMERS

EASE OF PROCESSING SPEARHEADS COMMERCIAL SUCCESS

A Report from
TECHNICAL INSIGHTS
John Wiley & Sons, Inc.

R-224: Conductive Polymers: Ease of Processing Spearheads Commercial Success

ISBN: 0-471-29861-1

Cover figure courtesy of Uniax Corporation.

ADDRESS INQUIRIES TO

Reports Group, Technical Insights
John Wiley & Sons, Inc.
32 North Dean Street
Englewood, NJ 07631-2807

Telephone:	201-568-4744
Fax:	201-568-8247
E-Mail:	reports@insights.com
URL:	http://www.insights.com

CONTENTS

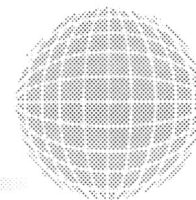

EXECUTIVE SUMMARY 1
  E.1   Objectives of Report 2
  E.2   Overview of Conductive Polymers. 2
  E.3   Scope and Methodology 3
  E.4   Summary of Major Findings 4

1. INTRODUCTION 7
  1.1   History of Conductive Polymers 7
  1.2   The Behavior of Materials 8
  1.3   Conductive Polymers. 9
  1.4   Pyrolytic Polymers 11
  1.5   Electronic Structure. 12
  1.6   Electrical Conductivity Models 13
  1.7   Materials Development 13
        1.7.1   Polyacetylene. 13
        1.7.2   Other Conductive Polymers 15
        1.7.3   Conductivity of Polyaniline. 15
        1.7.4   Processing Conductive Polymers 16
        1.7.5   Improving Conductivity 17
  1.8   Advantages Over Traditional Conductors. 19
        1.8.1   Environmentally Advantageous 20
        1.8.2   Ease of Processing. 24
  1.9   Commercialization. 26
        1.9.1   Future Products 27
        1.9.2   Issues and Challenges 30

2. APPLICATIONS AND MARKETS 33
  2.1   Market Overview 33
  2.2   Applications 34
        2.2.1   Light-Emitting Polymers. 34
        2.2.2   Cellular Phone Displays 37
        2.2.3   Thin-Film Transistors. 38
        2.2.4   Optical Switching 38
        2.2.5   EMI Shielding. 39
        2.2.6   Anticorrosion Coatings. 40
        2.2.7   Welding Plastics 41
        2.2.8   Conductive Textiles and Fabric 42
        2.2.9   Antistatic Films 43

2.2.10 Lithography and Printed Circuit Boards 43

2.2.11 Conductive Adhesives 45

2.2.12 Chemical Sensors 47

2.2.13 Electronic Noses 48

2.2.14 Batteries 50

2.2.15 Aerospace Applications 51

2.2.16 Molecular Conductors 53

2.2.17 Nanocomposites 54

2.2.18 Microrobotics 54

3. RESEARCH EFFORTS AND OPPORTUNITIES 57

3.1   Universities and Research Organizations 57

3.1.1   University of California, Berkley 57

3.1.2   University of Florida 57

3.1.3   Georgia Institute of Technology 58

3.1.4   Johns Hopkins University 60

3.1.5   NASA Lewis Research Center 61

3.1.6   Ohio State University 61

3.1.7   University of Michigan 64

3.1.8   University of Rhode Island 65

3.1.9   National University of Singapore 66

3.1.10   South Bank University London 66

3.1.11   University of Texas at Arlington 67

3.1.12   TNO University of Industrial Technology 69

3.1.13   University of Tsukuba 69

3.1.14   University of Utah 70

3.2   Companies 70

3.2.1   Abtech Scientific Inc. 70

3.2.2   Americhem Inc 71

3.2.3   AromaScan Inc. 72

3.2.4   Bayer Corp. 72

3.2.5   Cambridge Display Technology Ltd. 72

3.2.6   DSM BV 75

3.2.7   Hughes Aircraft Co. Research Laboratories 76

3.2.8   IBM 76

3.2.9   Lucent Technologies 77

3.2.10   Milliken Research Corp. 77

3.2.11   Monsanto Co. 78

3.2.12   NEC Corp. 78

3.2.13   Neotronics Scientific Inc. 79

3.2.14   Neste Oy Chemicals 80

3.2.15   Ormecon Chemie GmbH & Co. 80

3.2.16   Panasonic Industrial Corp. 81

3.2.17   Philips Components BV 82

3.2.18   Strategic Diagnostics Inc. 82

3.2.19   Uniax Corp. 82

APPENDICES
A. Recent U.S. Patents 85
B. Resources and Bibliography 89

FIGURES
1.1 Repeat units of several electronic polymers 10
1.2 Schematic illustrations of (a) 50% sulfonated and
    (b) 100% sulfonated polyanilines (self-doped forms) 16
3.1 Light-emitting device 83

EXECUTIVE SUMMARY

Two decades ago, scientists developed the first intrinsic electrically conductive polymer, doped polyacetylene. Although at first these polymers were not processable or air-stable, in the years since this 1977 discovery, researchers have developed conductive polymers that can be processed into powders, films, and fibers from a variety of solvents. They also are air-stable. Different versions of these polymers can be blended into other polymers to form electrically conductive blends.

Scientists have made a number of different polymers conductive in recent years. They have used such materials as polyaniline, polythiophene, polypyrrole, and polyacetylene. A variety of applications has emerged in recent years for these materials, especially for the polyanilines. These include coatings and blends for electrostatic dissipation, shielding from electromagnetic interference (EMI), electromagnetic radiation absorbers for welding plastics, conductive layers for light-emitting polymers, and anticorrosion coatings for use on iron, steel, and other metals.

The key attraction of these polymers over traditional conductive materials, including metals, is their ease of processing and their robustness. Devices incorporating conductive polymers require a balance of conductivity, processability, and stability. In recent years, researchers have been able to optimize all three properties simultaneously.

Among the most exciting applications is the use of conductive polymers in light-emitting devices (LEDs), replacing silicon as the substrate material found in traditional LEDs in clock radios, appliance and instrument read-outs, automotive dashboard displays, and aircraft cockpit displays.

A lab curiosity for many years, conductive polymers have begun to find use in consumer electronics and in antistatic textiles, some of which have military applications. Still, their commercial potential has only touched the tip of the iceberg. Their eventual use as a commercial anticorrosion coating sounds mundane, but probably holds the best chance among all applications for reaping revenues for companies involved in conductive polymer development.

E.1 OBJECTIVES OF THE REPORT

While many groups of investigators have performed much basic research on conductive polymers over the years, it is only within the last few years that some materials have reached the commercial marketplace. This report focuses on applied efforts aimed at developing conductive polymers for new industrial applications.

This report briefly reviews some of the already-established commercial uses for polymers, but devotes most of its analysis to research and development efforts at universities, government organizations, and companies worldwide where applications work is moving forward.

E.2 OVERVIEW OF CONDUCTIVE POLYMERS

Polymers usually are insulators. They are large organic molecules constructed out of smaller ones that are linked together in a long chain. They are insulators because their molecules do not contain any free electrons that carry current. To make these materials intrinsically electrically conductive, researchers have borrowed a technique used in making semiconductors—they dope the polymers.

Doping the polymers adds atoms that have electronic properties. The atoms added to the polymers either give off some of their spare electrons to the polymer's bonds, or they grab electrons from these bonds, in turn contributing positive charges to the polymer. Either way, the long-chain polymer becomes electrically unstable. When a voltage is applied to the polymer, the electrons will scamper over its length.

When they were first discovered, conductive polymers created visions of future transparent circuitry, electronic displays, and all-plastic, lightweight batteries. They were supposedly durable and inexpensive materials that also were easy to produce. Developers demonstrated how the material would be used in new displays and plastic transistors. But in the real world, one problem with the materials has been their instability in air. Research groups have been able to overcome this hurdle in some instances, and plastics are finding real world use as antistatic coatings and camouflage textiles.

Developers of applications as well as those performing basic research into conductive polymers do not completely comprehend exactly how the materials become conductive. But they believe that the purity of the polymers and the arrangement of the polymer chains play important roles. For example, by tweaking polyacetylene, researchers make the material conduct 50,000 amps/V/cm, an increase from an initial 60 amps/V/c.

Research is aimed at controlling the amount of conductivity that the polymers have. Although other polymers are more conductive, polyaniline is becoming the material of choice for many uses. Its properties are well known: it resembles the plastic used in 35 mm photographic film, it is easily produced and stable in air, and its electronic properties are easily customized. Just as important in terms of commercial economics, polyaniline is the most inexpensive of the conductive polymers. It also can be readily made into a variety of forms for different applications, such as thin films.

There are some limitations to the technology. When conductive polymers first evolved in the laboratory two decades ago, some foresaw the day when the material would replace copper wiring in many applications. This will not happen. Polyaniline conducts up to approximately 500 amps per volt per cm. However, copper conducts 100,000 times as much current and costs half as much.

Still, the electrical performance of polyaniline is more than adequate for some applications, such as braids on cable that are difficult to make. Braids make a coaxial cable flexible, enabling it to wind around a living room floor to reach a television set. Weaving copper wire into braids is a slow and very labor-intensive process. If workers could extrude braids of conductive polymer and lay the insulation over the cable in a single step, production speeds could increase severalfold, and costs would decline.

Among the major markets for conductive polymers is solid state electronics, including electrostatic dissipation. Electric charges cause extensive damage to electronic systems. Electrostatic damage to electronic equipment in the United States alone is greater than $15 billion.

E.3 Scope and Methodology

This report reviews advances in developing industrial applications for conductive polymers. It profiles R&D efforts at more than 30 organizations, including companies with products on the market or in development, and reviews research activities at universities and government organizations worldwide.

Included are research activities at the Georgia Institute of Technology, Atlanta, GA; Ohio State University, Columbus, OH; TNO Institute of Industrial Technology, Delft, Netherlands; University of California, Berkeley, CA; University of Michigan, Ann Arbor, MI; University of Rhode Island, Kingston, RI; and University of Tsukuba, Tsukuba, Japan.

The products and activities of companies discussed in this report include Bayer Corp. (Pittsburgh, PA); Cambridge Display Technology Ltd. (Cambridge, UK); Monsanto Co. (St. Louis, MO); Milliken Research Corp. (Spartanburg, SC); Philips Research, Eindhoven (Netherlands); and Uniax Inc. (Santa Barbara, CA).

Covered here are several industrial applications for conductive polymers, including their use in anticorrosion coatings, lighting displays, plastic batteries, the shielding of equipment from electromagnetic interference, textiles used for shielding humans and military aircraft, and welding materials.

The information presented in this report is based on an analysis and review of academic research papers and corporate product literature, as well as information found in the scientific and trade press. In addition, interviews were conducted with academic investigators and company executives.

E.4 SUMMARY OF MAJOR FINDINGS

A large variety of conductive polymer applications under development will find use in various industries. One industry that will reap benefits from conductive polymers is the electronics industry, where the polymers will be used as shielding against EMI and in electronic circuits themselves. In the detection and monitoring industries, the materials are already used in sensors in electronic noses that detect environmentally hazardous chemicals, factory emissions, and flavors or aromas in food products.

Also standing to benefit from conductive polymer technology are consumer electronics products in which the materials will be used in light-emitting polymers (LEPs) to power displays in clock radios, audio equipment, televisions, cellular telephones, home appliances, and industrial equipment. Even though some applications for conductive polymers already are commercial, many more uses require research and development efforts.

The principal advantages of polymers over conventional conductive materials are their potential ease of processing and relative robustness. Devices would need a balance of conductivity, processability, and stability, but, until recently researchers could not achieve all three properties simultaneously in conductive polymers.

One of the longest-established products incorporating conductive polymers is the fiber Contex, manufactured by Milliken since 1990. The fiber is coated with polypyrrole and can be woven to create an antistatic fabric. Milliken also markets camouflage netting based on Contex, which helps to conceal military targets from near-infrared and radar detection. Other products, either under development or on the market, include a Contex wrist rest for computer operators that helps dissipate static charge near computers; antistatic protective films and bags based on polythiophene for packaging electronic items sensitive to static discharge; and capacitors based on polypyrrole.

Despite all the promise that conductive polymers hold, researchers face several obstacles. The idea of using polymers as straight analogs of silicon in transistor devices has been difficult to achieve. Although these polymers can be viewed as semiconductors and metals in conventional band theory, in practice the conductivity is not band line but rather a "hopping" conductivity. The carriers move in a series of jumps, so that overall mobility is quite low. This limits the speed of response of the device. Moreover, silicon is already the material of choice in many applications.

However, there are still opportunities for relatively low-cost devices insensitive to mechanical deformations. Scientists at Lucent Technologies (Murray Hill, NJ), are developing thin-film transistors from conductive polymers that might be used on product packaging to replace bar codes. These new plastic transistors would contain pricing and other important merchandise data. The data would be retrieved by a scanner at the check-out aisle. Silicon-based transistors would not be as flexible as plastic transistors and would break on the packaging.

Meanwhile, scientists at Philips Research (Eindhoven, Netherlands) also have been investigating the potential of polymers in field-effect transistors. While some high-quality devices have emerged, there have been problems with the reproducibility of devices.

Scientists have made field-effect transistors from polythiophene using printing techniques. Rolling, bending, and twisting did not affect the transistor's electrical characteristics. Others have developed a water-soluble conductive polymer based on polyaniline. The materials can be easily applied to various surfaces, and they are curable. Exposing the polymer to radiation results in its crosslinking, making the material insoluble but still conductive.

These types of polymers might find use as charge dissipators in scanning electron microscopy and electron beam lithography, replacing metals used to prevent an unwanted buildup of charge. As well as being expensive to apply, the deposited metal is often difficult or impossible to remove, making the sample not usable. It is possible that new polyaniline coatings would be easily and cleanly removed by rinsing with water.

Scientists at the University of Durham in the UK have achieved some success by fluorinating thiophene, which yields a stable polymer with good solubility in common organic solvents.

They are continuing to optimize production of polyaniline.

The properties of conductive polymers make them candidates for use in sensors. Chemical data can be converted into a measurable electrical response. A very small change in the materials' redox composition brought about by small quantities of a range of chemicals can induce a large and rapid change in their electrical conductivity.

One company, Abtech Scientific Inc. (Yardley, PA), has developed a sensor based on polypyrrole on an array of microelectrodes. It has shown that a simple system using the catalyzed conversion of iodide to iodine can detect hydrogen peroxide. The polymer incorporates molybdenum together with iodide. Hydrogen peroxide converts the iodide to iodine, which oxidizes the polypyrrole, resulting in a measurable change in electrical conductivity.

This sensitivity to hydrogen peroxide enables Abtech to develop a range of enzyme biosensors. When immobilized glucose oxidase is incorporated into this transducer system, it acts as a glucose-sensitive biosensor. The enzyme-catalyzed oxidation of the glucose produces peroxide as a byproduct. The system also works with d-amino acid oxidase, as a d-amino acid detector, and with lactate oxidase, for detecting lactate. The company is also marketing a polymer-based system to monitor sulfate-reducing *Desulfovibrio* bacteria, which have been implicated in corrosion and biofouling.

At the Hughes Aircraft Co. Research Laboratories (Malibu, CA), scientists have used conductive polymers to detect and measure changes in industrially important nonpolar media. Information about the condition or quality of fluids used for cooking, industrial machining, vehicle lubrication, alternative fuels, or specialty coolants is vital to their efficient use. Data also help determine when they need replacing or replenishing. How their properties vary is often difficult to measure in situ because of the nonpolar nature of the media.

Investigators found that measuring the change in current flow in conductive polymers that are in contact with the medium is an excellent way to determine the condition of some fluids. Hydrocarbon-based lubricants, for example, degrade by oxidation, and their byproducts can be detected by conductive polymers.

Meanwhile others have taken commercially available 3-methylthiophene, *N*-methylpyrrole, aniline, and furan and polymerized them on a variety of substrates to create three kinds of polymer electrodes: an ion-selective electrode, a thin layer electrochemical detector for high-performance liquid chromatography, and a voltametric electrode. With the three classes of electrodes, researchers managed to detect inorganic ions and neurotransmitters, and they have analyzed a variety of organic and biological compounds.

There is also interest in using polymers as electromechanical actuators. It is possible to build an actuator by affixing two pieces of chemically synthesized polyaniline film mixed with HCl to opposite sides of cellophane tape. Applying a potential difference between the films causes the actuator to turn. ■

1. INTRODUCTION

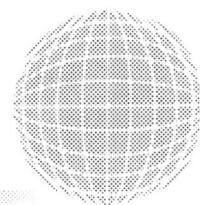

For the past 50 years, industries have increasingly used traditional polymer material systems as replacements for such structural materials as wood, ceramics, and metals. Reasons for this include polymers' high strength, light weight, adaptability, and ability to be produced at low temperatures.

In 1977, the first intrinsically electrically conductive organic polymer, doped polyacetylene, spurred great interest in developing these materials for industrial applications. Intrinsically conductive polymers are different from conductive polymers, which are mixes of a nonconductive polymer, normally used as an insulator, and a conductive material like metal or carbon black powder.

Although intrinsically conductive polymers were not initially processable or stable in air, the research community has come a long way in making conductive polymers that are. The latest in a long line of these compounds can be processed and made into films, coatings, fibers, and powders from a variety of solvents. It is possible to blend some versions of these polymers and mix them with conventional polymers to create electrically conductive blends. Some of the conductivities of the newer conductive polymers are in the range of those of silicon and copper.

Synthetic polymers, whose long molecules join tens or hundreds of identical structural units, have been used as insulators for almost 100 years, but it was not until the 1970s that the research community took note of new polymers that were able to conduct electrical current.

Polymers, as a family of materials, are relatively lightweight and are easy to process, two characteristics that account for the current great commercial interest in conductive polymers. Moreover, this conductive material is compatible mechanically and electrically with traditional conductors, such as copper, and traditional semiconductors, such as silicon.

Until the advent of conductive polymers, adding conductive fillers, such as carbon black and copper, to regular insulating polymers was the only way

1.1 HISTORY OF CONDUCTIVE POLYMERS

7

to create alternative conductors that were easy to process and had useful electrical conductivity. Unfortunately, there are problems with the use of fillers: surface corrosion, uneven mixing of materials, reduced mechanical properties, and incompatibility of a filler with its matrix.

1.2 THE BEHAVIOR OF MATERIALS

All materials can be broken down into three classes based on their behavior in an electric field: conductors, semiconductors, and insulators. To comprehend the differences among these behaviors, it is best to create a model for charge transfer within a solid. The basic aspects of band theory form the foundation of an understanding of the distinction among insulators, semiconductors, and traditional metallic conductors.

Atoms and molecules in isolation contain electron energy levels. Each electron within an atom may exist in one of various energy levels associated with that atom. Several scenarios are possible. One electron could occupy each energy level. In addition, two electrons of opposite spin also could occupy an energy level. On the other hand, the energy level could be empty. The electrons could move among these levels under the proper conditions.

When atoms assemble into an array to form a crystalline solid, the energy levels broaden into energy bands, which consist of very closely spaced energy levels. These levels are so closely spaced that they form a continuum. The bonds are a continuum of states, not just individual states of matter.

A solid material's electron behavior depends to a large extent on the occupation of these energy bands. If an electron within a solid is subjected to an electric field, a force will be placed on this electron. This causes the electron to accelerate, gaining energy and becoming part of an electric current within the solid material. For the electron to accelerate and gain energy, there must be an unoccupied, higher energy level that it can occupy. If there is no such level, the electron cannot accelerate, and there is no electrical conduction.

In the case of a metallic conductor, its valence band is completely filled, and its conduction band is partially filled. If an external electric field is applied, electrons have an unoccupied level above them. They accelerate in response to the applied field and create electrical conduction.

A semiconductor actually is an insulator with an energy gap that is smaller by an order of magnitude. Electrons near the top of the semiconductor's valence band have no unoccupied levels directly above them. The energy gap is much smaller. A number of electrons can be excited thermally across the gap into the conduction band. Once there, they become involved in conduction.

When an electron is excited across its gap, it leaves behind a hole in the valence band. This hole becomes a positively charged carrier. The dynamics of its motion are similar to those of electronic motion. Therefore, the most general form of electronic conduction in a semiconductor entails charge transfer by both electrons and holes.

Most polymers have molecular structures that are very disordered. Most are insulators, but some can be made conductive by altering their molecular structure or by doping them with impurities. Conductive polymers are composed of chains of alternating single and double bonds, aromatic rings, or heteroaromatic rings.

A major obstacle to the future development of new conductive polymers is a lack of understanding of how electrical current passes through them. All conductive polymers have one characteristic in common—they contain extended conjugated systems, where single and double bonds alternate along the polymer chain.

A basic objective among research teams is to understand the relationship between the chemical structure of a polymer's repeating unit and its electrical properties. Understanding this relationship would allow scientists to tailor the electronic and mechanical properties of the polymers at the molecular level.

Conductive polymers often are insulating materials until they are doped. Depending on the nature of the dopant, a partial oxidation or reduction of the polymer occurs as a result of the doping process. The resulting charge is delocalized along the chain of the polymer. In a pristine polymer, the electrons in the carbon p-orbitals comprise a complete valence band. There are no electrons in the conduction band. This type of polymer can at best function as a semiconductor and is often used as an insulating material.

When oxidative doping occurs, an electron exits the valence band, leaving a hole that is free to roam. If a reductive dopant is used, an electron from the dopant is transferred into the polymer conduction band, leaving it partially filled. Polyacetylene, one of the more extensively studied conductive polymers, is doped by either an oxidative or a reductive process. On the other hand, oxidation is more easily used with such heteroaromatic polymers as the polypyrroles and the polythiophenes (see Fig. 1.1).

These types of polymers have usually been difficult to handle because they have a tendency to be intractable and insoluble. After they are doped, the polymers have little environmental stability. They degrade when they are exposed to air, humidity, or temperature extremes. The conductivity

1.3 CONDUCTIVE POLYMERS

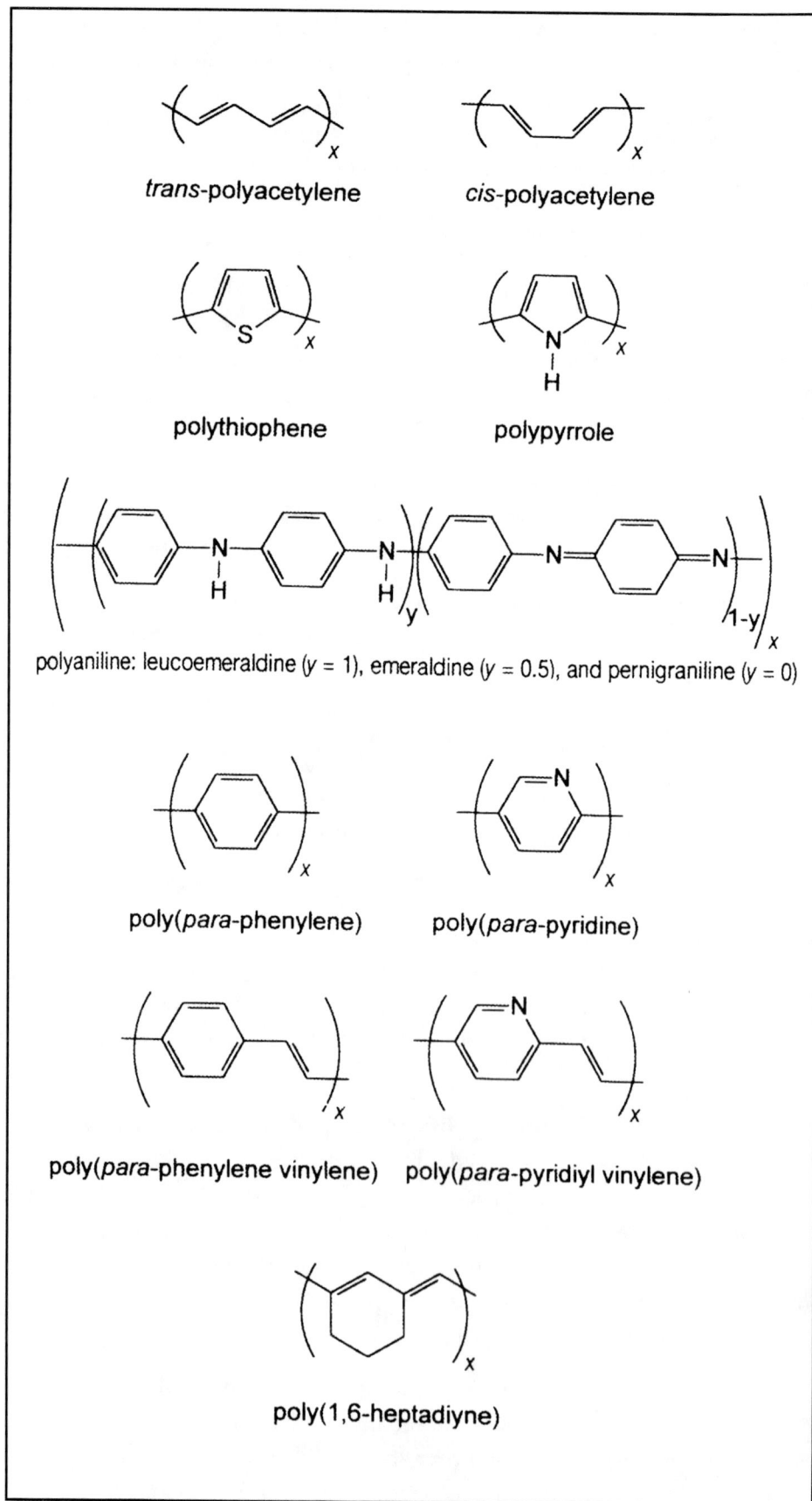

Permission of Materials Research Society.

Figure 1.1
Repeat units of several electronic polymers.

trans-polyacetylene

cis-polyacetylene

polythiophene

polypyrrole

polyaniline: leucoemeraldine ($y = 1$), emeraldine ($y = 0.5$), and pernigraniline ($y = 0$)

poly(para-phenylene)

poly(para-pyridine)

poly(para-phenylene vinylene)

poly(para-pyridiyl vinylene)

poly(1,6-heptadiyne)

appears to be sensitive to defects in the polymer chains, as well as to any impurities. These defects could cause scattering and localization of electron transport, rendering the polymers less conductive.

Polyaniline is one conductive polymer that is very environmentally stable. Scientists prepare this material by the chemical or electrochemical oxidative polymerization of aniline. The end result of this procedure is a material that is soluble in different organic solvents and aqueous acids. It also can be easily cast into films.

On another front, the solubility of such materials as polypyrrole and polythiophene can be increased by adding aliphatic side chains, which have few negative effects on a polymer's conductivity. Investigators have made conductive polymer composites by blending these materials with polystyrene in solution and casting films. The blending procedure may help overcome some problems, such as poor mechanical properties and environmental instability, inherent in these systems.

1.4 PYROLYTIC POLYMERS

Pyrolytic polymers are derived from the pyrolysis of high-char–yielding polymers. Good examples of these materials are the polyacrylonitriles, which have found use conductive polymers. Several high-char–yielding polymers proposed as matrix resins for carbon-carbon composites have become extremely conductive when pyrolyzed.

Among the more promising of this class of materials are the phthalonitriles. These are highly crosslinked polymers. Investigators make these resins conductive by pyrolyzing them at greater than 450 °C in an inert atmosphere. The conductivity of these materials is more stable than some doped polymers when exposed to air, water, or elevated temperatures. Unfortunately, these materials are inherently brittle, and scientists have found them difficult to process because of the high degree of crosslinking.

It is important to understand how charge-transfer systems function. Electrical conductivity in an organic solid occurs after the transfer of an electron, or charge, from an electron-rich donor molecule to an electron-poor acceptor molecule. This type of organic conductor, known as a charge-transfer salt, was initially discovered in the 1960s. The scientific community at the time believed these substances to be semiconductors. But in the early 1970s, researchers found that single crystals of some charge-transfer salts had conductivities similar to those found in metals.

Donor and acceptor molecules are often planar molecules. Charge-transfer occurs when an electron is transferred from the highest occupied molecular orbital of the donor to the lowest unoccupied molecular orbital of the acceptor. This transfer of electrons requires a large amount of overlap between the donor and acceptor. For this to occur, it must take place in a regular crystal structure, which charge-transfer salts possess.

In the solid state, charge-transfer salts crystallize into two segregated stacks. One stack contains all donor molecules, while the other contains all acceptor molecules. In each stack, the individual orbitals form a band of electronic states. The charge transfer of some of the electrons creates the partially filled bands needed for conduction to occur. The actual conduction occurs along the axes of these stacks.

It is possible to construct charge-transfer polymers by attaching the donor and acceptor moieties to a polymer chain in such a way that the required overlapping between the donor and acceptor molecules occurs. Most charge-transfer polymers are photoconductive—their conductivity is enhanced when they are irradiated with light at the correct wavelength.

Charge-transfer polymers have low levels of conductivity, which may be caused by a disorder that is inherent in these polymers. However, this low conductivity may also be caused by the types of donors and acceptors that have been incorporated into the polymers.

The charge-transfer salts that have the best conductivities may not always be incorporated into polymers. Charge-transfer polymers do have advantages. Because they are inherently conductive, unlike linear polyenes like polyacetylene, their conductivity is more environmentally stable.

1.5 ELECTRONIC STRUCTURE

The electronic ground state of electrically conductive polymers is that of an insulator. The conductivities of pristine electronic polymers are transformed from an insulating to a conductive state through the process of doping. Conductivity increases as the amount of doping increases.

The doping procedures differ from conventional ion implantation, which is used on semiconductors. Doping is carried out by exposing polymer films or powders to vapors or solutions of the dopant. In some instances, the polymer and the dopant are dissolved in the same solvent before the mix forms into a film or a powder.

In electronically conductive polymers, the atomic or molecular doping ions are positioned interstitially between the polymer chains and donate charges to or accept charges from the polymer backbone. The backbone and dopant ions form new three-dimensional structures. The structures vary. Different structures occur, depending on the levels of dopants used. Variations in processing also will cause different structures to develop.

The negative or positive charges initially added to a polymer's chain when the material is doped do not begin to fill the material's rigid conduction band or valence band, which would create a conductivity similar to a metal's. The strong coupling between electrons and phonons (vibrations) causes distortions of the bond lengths in the vicinity of the doped charges.

Because there is a large diversity in the properties of materials synthesized even by the same synthetic processes, it is important to perform structural, electrical transport, magnetic, and optical studies of the same materials. For example, the conductivities of some conductive polymers increase by several orders of magnitude when the pristine polymer is first doped.

The conductivity of one doped polymer, polyaniline, can vary significantly in magnitude and temperature dependence if it is doped in different solvents. The effects of a solvent and its vapors on the structure order and eventual electrical conductivity of intrinsically conductive polymers, especially the polyanilines, is called secondary doping.

The electrical conductivity of any conductive polymer can be altered over a wide range, depending on the amount of reactivity of the dopant. The doping process creates spinless charge carriers, called polarons and bipolarons, which exist at energy levels within the band gap. This is unlike the situation in metallic conductors, in which the charge carriers, the electrons, have spin. Conduction is thermally activated and depends exponentially on temperature, similar to that which occurs in semiconductors, but unlike what occurs in metals.

1.6 ELECTRICAL CONDUCTIVITY MODELS

Much research has been devoted to understanding the nature of the charge carriers in highly doped polymers. Although there is a large number of conduction electrons at the chemical potential in the highly doped state, the carriers may not be localized spatially so that they cannot participate in transport, except by hopping. Studies of some conductive polymer systems show that they contain modest amounts of crystallinity. Some regions of the materials are more ordered, while other areas are more disordered.

The percent of crystallinity varies from zero to 60% for polypyrroles and polyanilines. It can be greater than 80% for polyacetylenes. The chains in the disordered regions can be either relatively straight, tightly coiled, or intermediate in terms of disorder. Impurities and lattice defects in disordered material systems introduce backscattering of electrons. The electronic structure of a material system significantly depends on its degree of disorder.

1.7 MATERIALS DEVELOPMENT

Among the many polymers known to be conductive, polyacetylene, polyaniline, polypyrrole, polythiophene, poly(phenylenesulfide), and poly(phenylenevinylene) have been studied the most extensively. Polyaniline was the first of these to be commercialized.

1.7.1 POLYACETYLENE

The material that actually launched research into conductive polymers is polyacetylene. This conductive polymer is a simple macromolecule with significant electrical properties. It is complex physically, containing weakly coupled chains of carbon-hydrogen units arranged in a one-dimensional lattice.

Although polyacetylene was already known to exist as a powder, in 1974 scientists at the Tokyo Institute of Technology in Japan accidentally prepared it from acetylene as a silvery film using a Ziegler-Natta–type polymerization catalyst. Despite its appearance, the material was an insulator, not a conductor.

The Japanese scientists and colleagues at the University of Pennsylvania (Philadelphia, PA), and the University of California (Santa Barbara, CA), made the next developmental breakthrough in the field in the late 1970s. They discovered that partial oxidation using iodine or other reagents made polyacetylene films 10^9 more conductive. This conductive form of the material was called doped polyacetylene. The conductivity achieved by investigators was greater than that of any known polymer. But even though polyacetylene has high conductivity, it was not among the first conductive polymers to be commercialized because it was not stable in air or moisture.

Polyacetylene has two structural forms. The first, *trans* type is thermodynamically more stable. The polyacetylene made at the Tokyo Institute of Technology was the *cis* version, which converts to the *trans* structure when heated at greater than 150 °C. The actual doping process that turns ordinary polyacetylene into a good electrical conductor involves an ordinary oxidation process, called p-doping. This means that the material's electrical properties can be altered using chemical techniques. Scientists also have investigated using reductive doping (or n-doping) on the polymers. However, this approach makes polymers even more sensitive than undoped polymers.

Chemically, doped polyacetylene is a salt with oppositely charged ions, one of which is a good electrical conductor. Conductive polymers can be doped and then undoped (switched from the conductive to the insulating state) by applying an electrical potential. This electrical potential causes the dopant ions to diffuse in and out of the polymer. Thus, researchers are able to alter the material's properties in a controlled manner. This is a very appealing feature of conductive polymers when it comes to developing practical applications in which conductivity has to be precisely controlled.

Polyacetylene is the most crystalline of conductive polymers. The structure of undoped and iodine-doped polyacetylene has been studied extensively. Studies of the doping of *trans*-polyacetylene using alkali metals reveal that the packing of the polymer chains depends for the most part on the size of the alkali ions. Lithium-doped polyacetylene contains about the same volume as the undoped polymer. Thus, any volume expansion that takes place during the doping process is limited.

This limited expansion is a desirable feature for lithium-polyacetylene batteries. For example, if polyacetylene expanded, it could break a battery. In a rechargeable battery, when all of the polymer is converted from the p-doped to the neutral state, the cell is fully discharged. In order to recharge the cell, an opposite potential is applied to the electrodes.

Since the early 1980s, three other polymers—polypyrrole, polythiophene, and polyaniline—have been studied to a large extent to determine their conductivity. Around 1980, scientists at IBM (Yorktown Heights, NY) found that polypyrrole, which had been known only as an intractable black powder, could be made into a film by electrochemically oxidizing pyrrole in acetonitrile. Polypyrrole formed at the electrode's surface and could be peeled off as a flexible, dense, shiny blue-black film.

In the early 1980s, polythiophene was produced similarly in France and by IBM. The scientists made the material using the anodic oxidation of thiophene. Because this electrochemical technique makes it possible to control the oxidation potential of the polymerization process, it is possible to optimize the quality of the polymer. This technique forms polymers in the doped state as films that usually have favorable mechanical properties. It is widely used in synthesizing many conductive polymers.

Polypyrrole and polythiophene differ from polyacetylene in several ways. They are synthesized directly in the doped form and are extremely stable when exposed to air. But they are less conductive than polyacetylene. The oxidative polymerization of polypyrrole and polythiophene goes through very reactive radical cation intermediates that couple irregularly.

The first water-soluble polythiophenes were made in the late 1980s. The alkyl chains on thiophene were substituted with polar alkylsulfonate chains. The thiophene then was polymerized. These materials are similar to soaps: They contain both a hydrophobic and a hydrophilic portion. They have anions covalently attached to them and do not need an external dopant anion—they are self-doped (see Fig. 1.2).

Polyaniline consists of up to 1000 or more repeating units. This material exists in several oxidation states with varying electrical conductivities. Different compositions of polyaniline have different colors and electrical properties. One form, emeraldine salt, is electrically conductive. Scientists synthesize this version using electrochemical or chemical oxidation of aniline in aqueous acidic media and common oxidants. Some have made mechanically flexible, dark blue films of conductive polyaniline by the protonic doping of emeraldine films cast from methylpyrrolidinone solutions. Protonic doping, performed by dipping the films in acid or by passing a gaseous acid over them, protonates the imine nitrogen atoms in the backbone of the polymer. The conductive emeraldine salt becomes the insulating emeraldine base when treated with aqueous alkali.

Before the early 1980s, scientists performed little research on polyaniline, although they knew that the material's conductivity increases by orders of

1.7.2 OTHER CONDUCTIVE POLYMERS

1.7.3 CONDUCTIVITY OF POLYANILINE

15

Figure 1.2

Schematic illustrations of (a) 50% sulfonated and (b) 100% sulfonated polyanilines (self-doped forms).

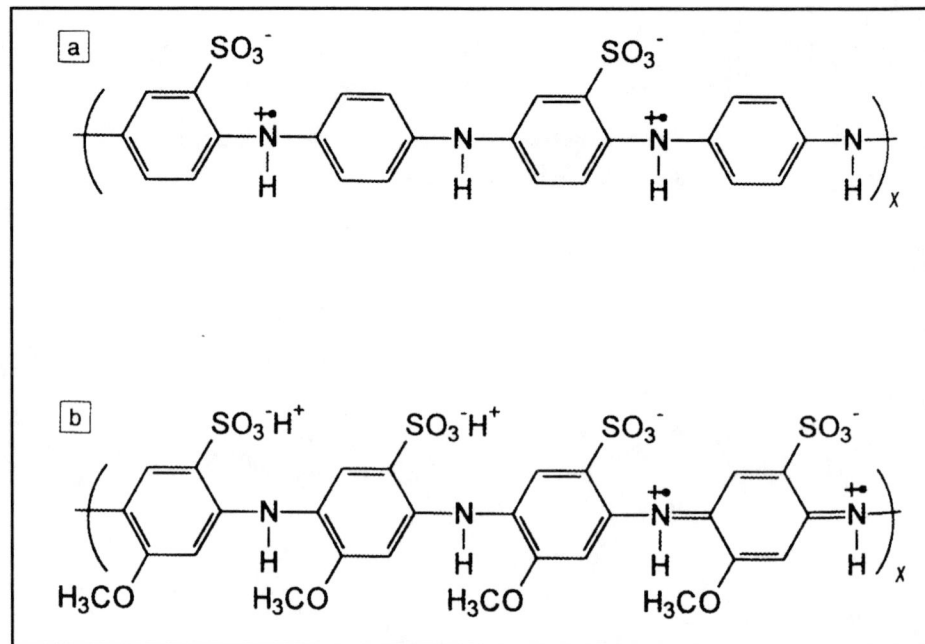

magnitude as the pH of the acid with which it is doped decreases. They also realized that the polymer could serve as an electrode material for batteries.

During the 1980s, polyaniline was extensively studied in terms of its structural, physical, and electrical characteristics. Investigators determined the crystal structures of several pure oligoanilines (parts of polyaniline). Polyanilines are amorphous, but both the emeraldine base and emeraldine salt are partially crystalline in form.

Polyaniline differs from other conductive polymers in its emeraldine form in that partial oxidation or reduction is not needed to dope it. Researchers create a highly conductive material by the protonation of the imine nitrogen atoms in the backbone of the emeraldine base. This protonic doping does not alter the number of electrons in the polymer. The pH of the solution used to treat the polymer determines its conductivity.

1.7.4 PROCESSING CONDUCTIVE POLYMERS

Polyacetylenes, polypyrroles, and polythiophenes are difficult to process because they do not dissolve or melt. The amorphous nature of such materials makes it difficult to determine their exact structures. This, in turn, makes it difficult for scientists to improve upon their properties so that they can be processed with less difficulty.

Still, new processing pathways have been charted. Scientists at the University of Durham in England have synthesized polyacetylene from a soluble precursor polymer. They formed an undoped conjugated polymer after thermally treating the material. In the next step, they doped the polymer to a highly conductive state. During thermal treatment, a stable volatile organic molecule was ejected from the material's structure to leave poly-acetylene film. Others have made polyacetylene using a different tech-nique. They used the metathesis polymerization of cyclo-octatetraene, catalyzed by a titanium alkylidene complex.

One major advance in electrical properties occurred in the late 1980s when researchers harnessed a new polymerization method that entailed aging catalysts in silicone oil. When doped, the material had a conductiv-ity of about 0.25% that of copper when measured by volume. Its conduc-tivity was greater than that of copper when measured by weight.

The undoped, nonconductive versions of many conductive polymers are wide band gap semiconductors. They have a large energy spacing between the valence band and the conduction band. One goal of researchers in this field is to synthesize polymers that have a very small band gap. These materials would be intrinsically conductive and would not require any doping. Their high levels of conductivity would be caused by thermally exciting electrons from the valence band to the conduction band at room temperature. These types of materials would be doped by partial oxida-tion or reduction to have conductivities similar to those of metals. In this manner, it might be possible to create composites with excellent mechani-cal or electrical properties.

1.7.5 IMPROVING CONDUCTIVITY

The electrical conductivity of many conductive polymers has been about the same as for many inorganic semiconductors. These semiconductors have fewer carriers than conductive polymers. Unfortunately, the electron mobility of conductive polymers is limited, so if the conductivity of these materials is to be increased, there must be an increase in mobility. Higher mobilities will result from the development of more crystalline, better ori-ented, defect-free materials.

Researchers have used several approaches to align the chains more per-fectly in conductive polymers, in turn achieving higher mobilities. One technique entails polymerizing monomers in liquid crystal solvents. Poly-merizing polyacetylene in a liquid crystal solvent creates highly oriented films with good conductivities after they are doped.

Others have made highly conductive films of water-soluble polythiophenes by passing a solution of the material through a magnetic field while removing the solvent. The resultant polymer's chains have a high degree of alignment.

Yet another avenue to take when aligning polymers so that they are more con-ductive is to stretch the polymer into films or fibers. Researchers found that polypyrrole films prepared at -20 °C could be stretched to twice their original length and that such films were 20 times more conductive than the unstretched films. It also is possible to create polymer films that have good mechanical and electrical properties. One such material is *trans*-polyacetylene.

On another front, two polymers that can be made in a highly oriented form are poly(phenylenevinylene) and poly(thienylenevinylene). These polymers traditionally are insoluble and amorphous, but they can be made highly oriented if they are prepared by heating highly oriented films and fibers of a processable, soluble nonconjugated precursor polymer.

In order to align the molecules of these polymers and achieve crystallization, several research teams have encapsulated conjugated polymeric chains inside crystalline inorganic host materials. The inorganic host offers an environment that will favor polymer ordering in the polymer's chains. Moreover, because the polymerization reaction occurs in the host, fewer defects should result within the chains. Other scientific groups have synthesized very narrow polypyrrole and polythiophene fibers by growing them electrochemically within the pores of microporous membranes, resulting in high levels of conductivity.

Other conductive polymers have been developed in addition to those discussed in this chapter. For example, the LEP display technology under development at Uniax (Santa Barbara, CA) is based on one of several 2,5-substituted chemical derivatives of the polymer poly(*p*-phenylene vinylene) (PPV). PPV is used as the active layer in these displays.

The generic structure of these materials is based on a conjugated polymer backbone containing alternating 1,4-phenyl and *trans*-vinylene moieties. These polymers usually have at least 100 repeat units. Varying the side groups has a dramatic effect on the physical and electro-optical properties of the resulting substituted PPVs. Scientists at Uniax are currently focusing on a number of these derivatives, of which the most interesting is the 2-methoxy,5-(2'-ethyl)hexyloxy derivative denoted MEH-PPV.

This polymer can be cast into high-quality thin films with a large internal quantum efficiency of conversion of injected electron-hole pairs to red-orange light. Substitution of other side groups or the use of different backbone structures leads to electroluminescent emission by the polymer, which makes it ideal for light-emitting displays found in consumer electronics.

It is possible to create different colors ranging throughout the visible spectrum. The ability to make various derivatives of the polymer simplifies device fabrication. It is possible to directly cast the polymer layer from solution using process steps like those used to deposit a polymer layer in active matrix liquid crystal displays or the insulating polyimide layers in multichip modules.

The structure of an LED made using this polymer is not complex. Fabricating the device involves four steps. First, a patterned array of optically transparent conductive electrodes is formed on a suitable transparent substrate. The typical substrate might be a sheet of either a commodity plastic film (polyethylene terephthalate) or glass coated with ITO (indium tin oxide). Both of these materials are readily available commercially. ITO makes a good hole-injecting contact to semiconductive polymers. Continuous layers of ITO, supplied by a commercial vendor, can subsequently be patterned using standard photolithographic processes.

In the next step, a thin layer of luminescent semiconductive polymer is spin-cast onto the prepared substrate coated with a transparent electrically conductive film. The spin coating process, which is now in commercial use for substrates with diagonal measurements up to 24 inches, has undergone intensive development over the past decade by the makers of flat panel displays. The technique produces optical quality submicrometer-scale films with uniform thickness.

In the third step, a functional metal, such as calcium or magnesium, is deposited on top of the spin-cast polymer layer and array of electron-injecting strip electrodes. This is done under high vacuum (about 4×10^{-7} Torr). A protective overcoat of a relatively nonreactive metal, such as aluminum, is usually applied over these strips before the final encapsulation step. Then a protective back cover is fixed to the open surface of the device using an adhesive. The cover is cut smaller than the substrate, so that the row and column electrodes can extend beyond the cover and a connection can be made to external drive circuitry.

When a sufficiently large bias voltage is applied between a pair of row and column electrode strips, light is emitted through the transparent substrate from the luminescent polymer sandwiched between the metal and ITO contacts at their intersection.

1.8 ADVANTAGES OVER TRADITIONAL CONDUCTORS

Many conductive polymers offer advantages over other electrical conductors. The electrical conductivity of the polymers can be controlled over a wide range. A variety of commodity polymers, thermoplastics, and elastomers can be used to make conductive blends and dispersions. Scientists can turn electrical conductivity on or off (and back again) with an acid-base reaction. It is possible to make thin, flexible transparent films and coatings electrically conductive. The metallic surfaces of bridges and other structures can be passivated, and made to resist corrosion, by coating them with conductive polymers. This creates a protective oxide layer.

A good way to demonstrate the advantages of using plastics as conductors of electricity is to briefly review potential applications for the material. In the electronics area, contemporary protective packaging depends on ionic salts or resins filled with metals or carbon. But this packaging has its shortcomings. The conductivities of ionic materials often are low and unstable. Metal is heavy and expensive. Carbon is a potential contamination hazard because small pieces can slough off when equipment is shipped.

However, polymers are less difficult to handle and should be able to dissipate electrostatic charges more efficiently. In addition, polyaniline coatings are very transparent. IBM already is marketing a polyaniline, called PanAquas, for this type of protective application.

On another front, the dissipative abilities of conductive polymers make them strong candidates for electromagnetic shielding applications. This protection is necessary to keep electrical signals from overlapping. This is why people are not allowed to use portable electronics during the takeoff and landing of aircraft.

Conductive polymers guard against these emissions and signals when they are incorporated into the plastic coverings and casings of electronics equipment. The problem with traditional screening materials is that they utilize impregnated pieces of carbon or metal, which could harm the mechanical properties of the base material at any point at which it bends. Conductive polymers could be blended with other materials, such as nylon, to reduce their cost of usage, while still performing a shielding function. (Note: The shielding ability of conductive polymers would not help to shield consumers from potential negative health effects of overhead power lines. The frequencies of these fields are much lower than that which the polymers block.)

1.8.1 ENVIRONMENTALLY ADVANTAGEOUS

A polymer often must be processed with organic solvents. If water-soluble polyaniline (e.g., Pan Aquas) could be made more conductive, it could replace the lead-based solder that connects electronic components on a substrate. In some instances, manufacturers must remove lead-containing material from discarded printed circuit boards, which is a difficult task.

Also environmentally advantageous would be an electronics component made of polymers. Several research teams have been developing an all-plastic transistor using conductive polymers. However, such a transistor could not compete in all applications with silicon. For example, computer electronics made with polymers would operate at less than one-thousandth the speed of those using crystalline silicon.

Still, there is one application that does not require extremely fast electronic circuitry: video displays. Amorphous silicon is used in video display circuitry because it is less expensive to process than crystalline silicon. Moreover, the amorphous material can be applied to different substrates, including a glass surface.

Several research groups have made conductive polymer transistors that operate at approximately the same speed as circuits that are fabricated from amorphous silicon. The required video performance could be within reach.

An organic semiconductive transistor would offer many advantages to makers of LCDs. LCDs are the focus of much research into flat panel displays. Existing screens seal liquid crystals, which are made from organic substances, between two plates of glass. A fluorescent light illuminates the crystals from behind. In passive displays, the pixels, which contain the liquid crystals, are controlled by voltages applied all along the rows and columns of the display. In active-matrix displays, which have better contrast and resolution, each pixel is controlled by a thin film transistor.

The issue of cost is important here. A 20-inch, full-color active-matrix display contains more than 2 million pixels. It just takes a few malfunctioning ones to ruin an image. A high failure rate results in an increase in the price of the displays that reach the market.

Organic-based circuits comprised of conductive polymers could ease this cost and lower consumer price tags. They should be easier to make, especially in large sizes. Circuitry can be fabricated at lower temperatures. Such circuits would be less sensitive to impurities while they are being fabricated. Fewer impurity-caused defects should also lower production costs.

Moreover, researchers should be able to make new types of displays using conductive polymers. Producers should be able to tailor and fine-tune the properties of the polymers, controlling their flexibility and transparency. Case in point: See-through electronics might make it possible to produce a direct-view, heads-up display on windshields and helmets, eliminating any need to reflect imagery onto a viewing glass.

The LCD display is typically the most expensive, bulkiest, heaviest, and most power-hungry component in a cellular telephone. Most models available today display between 8 and 20 alphanumeric characters in one to four lines. They may also include icons indicating status of functions, such as the state of the unit's battery charge or its signal strength.

Today, 80% of cellular telephones use LCDs for displays. A major disadvantage of LCDs is that they need to be illuminated by backlight for viewing in poorly lighted environments, which consumes a great deal of power. The remaining 20% of the cellular telephone display market is composed of intelligent displays based on matrix-addressable arrays of inorganic LEDs. These comparatively expensive displays consume even more power than LCDs, but they can be used in low-light situations since they are emissive. Consumers find them attractive, and they are widely used in the higher-end telephones that enjoy the largest profit margins.

LED display modules face a seemingly insurmountable challenge in the near future. The clash between the demands for increased information density and lower cost is expected to price them out of the market in the next 5 years. Each dot added to an LED dot-matrix display adds a fixed increment of cost to the display at a time when manufacturers are aggressively trying to reduce cost. This situation provides a lucrative opportunity for polymer light-emitting devices.

Early polymer LEDs will be fabricated on and packaged in glass. With further development, the production of polymer LEDs on plastic substrates will be thinner and lighter and have greater mechanical flexibility and resistance to fracture.

For portable handheld applications, electroluminescent conductive polymer displays offer advantages over the two primary competing display technologies: LCDs with backlights and inorganic LEDs. Polymer displays consume less power. Unlike LCD displays, polymer displays have a good contrast ratio (the ability to view the display in strong room light). They are visible from wide viewing angles, whereas LCD displays have a limited viewing angle because of the nature of the light passed through or reflected through the display. Those with laptop computers or LCD display-based cellular telephones will readily see the advantage of an improved viewing angle. Also, the edge definition of the light generated from a polymeric display is better than that from an LCD, increasing both readability and resolution.

LCD displays also have an inherent response time limitation known as latency. This deficiency is exacerbated at lower temperatures, and the LCD display function stops at below freezing temperatures without supplemental heating. Conductive polymer displays have extremely fast response times and have no limitations involving display time or response time at lower temperatures.

Polymer displays can achieve higher resolution than inorganic LED displays. Due to the shape of the light beam emitted by LEDs, more dense pixel arrays with less spacing between individual LEDs are limited by pixel bloom. This is because adjacent elements that are not supposed to be lighted are illuminated by a nearby pixel. As a result, the viewer sees more apparently lighted pixels and the image becomes blurred. Uniax is developing displays for which the shape of the emitted light and the method of device fabrication eliminate pixel bloom. Thus, polymer devices generate sharp light output, or resolution, and can be fabricated cost-effectively in high-pixel-density configurations required for displaying symbol graphics, such as Asian characters.

In another area, conductive polymers are also considered a strong candidate for replacing conventional tin lead solders. Actually, the polymers would be used in a blend of adhesives. These conductive adhesives potentially could surpass the reliability of tin lead solders used in the electronics industry in surface mount manufacturing.

Conductive adhesives could be tailored to meet many requirements for use in printed circuit boards, for example. The materials could be modified to obtain the optimum conductivity, fatigue life, and strength as well as mechanical and electrical properties.

Many research groups are developing conductive anticorrosion polymer films. In this area, conducting polymers appear to have several advantages over currently used materials. When coating and protecting high-strength, lightweight aluminum alloy structures, such as aircraft, a urethane top coat, a primer coat, and a chromate conversion coating on the metallic surface are often applied. A major problem facing this industry, however, is the known carcinogenic properties of common chromate-based anticorrosion coatings. As a result, there is considerable interest in finding a suitable and more environmentally friendly substitute. Toward this goal, scientists at the University of Rhode Island (Kingston, RI) have developed a new polyaniline to replace the chromate conversion coatings on aluminum alloys.

The concept of using conductive polymer films and coatings to protect steel, aluminum, and other metals from corrosion was first suggested in the mid-1980s. It appears that conductive polymers protect the surface of steel as the result of a passive oxide layer formed by an anodization process. Other research shows that a polymer-coated surface scratched to expose underlying metal still offers corrosion protection through an electroactive galvanic coupling mechanism.

Past research on steel also indicates that a conductive polymer coating performed better in an acidic environment than current protective barriers. This success appears to be a result of the polymer's conductive stability in different pH and solvent environments. The electroactive nature of the conductive form of the polymer is needed to offer corrosion protection to the metallic surface.

Polyaniline behaves like a noble metal because its redox potential is close to that of silver. The polymer ennobles the surface of conventional metals, it transforms the surface of the metal to be protected into a thin but dense metal oxide layer, and it passivates metals, which no commercial coatings can.

During a complex reaction mechanism, iron (or in stainless steel) is converted to Fe_2O_3 in a process similar to the passivation of aluminum to Al_2O_3 by air. Researchers assume that such a reaction will also occur with other metals like copper, aluminum, or zinc, all of which are ennobled by electrically conductive polyaniline.

A polyaniline protective coating shifts the corrosion potential up to 800 mV for iron and steel and more than 2 V for copper. This leads to a dramatic decrease of the corrosion velocity under certain corrosion environments. Such a coating can find use as a primer or as a concentrate for the development and production of primers. A polyaniline corrosion coating system might include an ennobling primer and a sealing top coat.

On another front, conductive polymer-based batteries have advantages over conventional units. Unlike nickel-cadmium rechargeable batteries, all-polymer batteries, in which both the electrodes and electrolyte are polymers, do not contain heavy metals that could potentially contaminate soil and water. The polymer batteries also contain no liquids that might leak and pose safety hazards.

All-plastic batteries, such as the unit developed at Johns Hopkins University (Baltimore, MD), operate efficiently in extreme heat or cold, whereas extreme temperatures have negative effects on other battery materials. A car battery, for example, does not start easily when it is very cold outside. In contrast, the properties of a conductive polymer battery and its ability to function do not change with the temperature conditions—they have functioned in a -40-45 °C range in the laboratory. All-plastic batteries can be recharged hundreds of times and continue to function well.

In addition, such a power cell's thin sandwich design makes it very adaptable. The anode and the cathode are made of thin foil-like plastic sheets. The electrolyte is a polymer gel film that is located between the electrodes, holding the battery together. There is a size advantage as well. The cell could be as thin as a business card, although more power-intensive applications probably would require larger units.

Still, the thin, flat design would enable battery users to cut a cell to fit a specific space or configuration. Researchers conceive of a polymer battery in the form of a large thin sheet that could occupy an entire wall. Or the battery might be rolled up into a tube, like AA batteries.

Because polymer batteries can take on various configurations, they could be used in space satellites. In this application, sheets of batteries could be slipped into crevices and curves without adding much extra weight to a spacecraft. If connected to solar cells, they could be recharged by the sun's rays while the satellite is in orbit.

1.8.2 EASE OF PROCESSING

Among the advantages of conductive polymers over traditional conductors is their ease of processing. They can be spin-coated or even printed onto a substrate, unlike some metals. Organic materials also are very compatible with several different types of substrates.

Lucent Technologies is attempting to harness some of these advantages in the development of plastic transistors. When conventional transistors that employ silicon or another inorganic material are made, very high temperatures (about 1000 °C), and a high-vacuum environment are needed. But if a conductive polymer is used, this material can be printed onto a substrate, just as ink is printed onto paper, without the need for these extreme processing environments.

Screen printing is an environmentally friendly way to produce electronic circuitry and interconnections. Researchers form patterns in a single step with this technique. With a fine pitch of printed lines, the printing process can reduce the amount of time and costs associated with using photolithography. It should be possible to use the printing technique with a liquid-processable (or soluble) organic semiconductor to produce less expensive large-area displays or other electronics that have flexible plastic substrates for the display or the storage of data.

Conductive organic materials offer advantages if they are used as the light sources in displays, and not just in the circuitry. Currently, LEDs are made from gallium arsenide, which is an inorganic semiconductor. Two layers, each doped to have different electrical characteristics, are interconnected and function as positive and negative electrodes. When electricity passes through the materials, one electrode gives off the electrons. The other gives off positively charged holes, which are the spaces that the electrons usually occupy. The negative and positive charges meet at the conjunction of the layers, where they combine and emit light. The dopant's and semiconductor's properties determine the light's color—those producing green and red are the easiest to make.

Conductive polymer–based LEDs will make it less expensive to manufacture these displays. They will allow companies to reduce the number of contacts and interconnections in these devices. Traditional LEDs must be joined together to be used in displays found in clocks and kitchen appliances. It is not possible to make a LED larger than the grown gallium arsenide wafers—usually about 6 inches diagonally.

To create larger displays, manufacturers must individually mount and wire the LEDs. This is a difficult job because one letter of average size in a display requires 35 LEDs for it to appear. On the other hand, organic films can be laid over practically unlimited areas. The starting materials for polymers are less expensive than those for conventional LED materials.

1.9 COMMERCIALIZATION

Much of the commercial interest in conductive polymers centers around their use as corrosion-resistant coatings, coatings for electrostatic dissipation, and in fabrics. Several firms already have or will soon commercialize products that address these applications. Zipperling Kessler GmbH (Ahrensburg, Germany) has formed Ormecon AG to focus its efforts on these areas.

Monsanto Co. (St. Louis, MO) has acquired the polyaniline business of AlliedSignal (Morristown, NJ). Monsanto also purchased Allied's patents. In addition to the acquired polyaniline technology, called Versicon, Monsanto has some polyaniline technology of its own.

Meanwhile, Bayer AG (Leverkusen, Germany), Neste Oy (Espoo, Finland), and DSM BV (Geleen, Netherlands), also have joined the market fray. DSM has commercialized Conquest, a waterborne polyurethane dispersion of polypyrrole, for use in antistatic coatings and in coatings that combat EMI (electromagnetic interference) in electronics equipment. DSM also is developing a formula for anticorrosion applications.

Neste has developed Panipol in conjunction with Uniax. This technology is aimed at fiber spinning, injection molding, and film extrusion. Bayer is marketing antistatic coatings of a polyethylene and is investigating electronics applications too.

Sensors that use conductive polymers also are on the market in electronic or artificial noses used to detect aromas and flavors in foods or chemical compounds in the environment. Neotronics Scientific Ltd. (Essex, UK) has introduced such an electronic nose. The unit utilizes a polypyrrole sensor to determine subtle distinctions among complex odors. Aroma-Scan Inc. (Hollis, NH) is marketing an electronic nose that uses polypyrrole-, polyaniline-, or polythiophene-based sensors.

Meanwhile Americhem (Concord, NC) has developed and commercialized transparent conductive coatings and processable compounds that could be used for extrusion and molding. The company has used proprietary techniques, although its products are based on Versicon, a doped polyaniline. Americhem's conductive coatings are liquid dispersions of Versicon in specific film-forming matrices. The coatings are intended to adhere well to various plastics, glass, and metal. Americhem also has commercialized transparent conductive coatings for electronics packaging applications, such as EMI shielding.

In addition, Milliken Research (Spartanburg, SC), a unit of the Milliken and Co. textile company, is marketing Contex conductive textiles. Milliken is able to coat each fiber of a textile with up to 5% of doped polypyrrole. The substrates used include polyester, nylon, glass, and Kevlar. The materials are stable in air and are degraded only by strong oxidants and alkaline solutions. Company scientists found that encapsulating the polymer coatings and matrices keeps oxygen and ions out of the material, extending the product's longevity.

A number of electronic components on the market contain conductive polymers. Matsushita (Osaka, Japan) is marketing its SP Cap, which contains polypyrrole and aluminum. Sanyo Electric Co. (Osaka, Japan) is selling the OS condenser, which is made from tetracyano quinodimethane. NEC Corp. (Tokyo, Japan) sells a polypyrrole-based Neocapacitor, whose lower impedance and higher frequency may counter its price tag, which is about three times higher than that for conventional condensers. Also, Kanebo Ltd. (Osaka, Japan) is marketing a polyacene-based capacitor that is finding use in cellular phones, memory cards, and some watches.

1.9.1 FUTURE PRODUCTS

The number of potential products in which conductive polymers will be used is nearly unlimited, ranging from plastic batteries to compact disks. All-polymer batteries, injection-molded antistatic products, printed circuit boards, electrochromic smart windows, electrochromic automotive rear vision systems, paint primers, antistatic flooring and work surfaces, and conductive pipes for mining explosives are among the future possibilities for this technology.

The electronic display–based product segment itself offers a number of possibilities. Low-level backlighting products for LCD displays, toys, watch faces, and advertising are among the potential uses. Although some of these applications require plastic substrates and packaging materials so that they can be made very thin, these requirements can be met using devices built with conductive polymers placed on glass substrates.

Many LCD displays currently rely on ambient light for operation. However, these types of displays, which need to be seen in low-level lighting or at night, require an auxiliary source of light. Because the reflector at the back of the simplest LCD display is partially transmissive, a backlight can be used for low-level light conditions. This type of display module is currently incorporated in many portable and handheld applications, including cellular phones, pagers, and portable radios. Low power usage and uniform light output are among the desirable features of polymer LEDs that make them attractive candidates for these applications.

In these products, the LCD module itself is made from glass, so a glass substrate polymer backlight is compatible and thinner than edge or backlighted inorganic LED lighting fixtures. The typical display area of these devices is about 1.75 x 0.5–1 inch. These dimensions can be accommodated by current polymer LED technology. Polymer LED backlights also should find use in other hand-held devices that have larger display areas, such as personal digital and specialty computers.

In addition to handheld equipment, polymer backlight and display technology could be used in a variety of other consumer products, including household appliances, home audio equipment, some wired telephones, short-wave radios, and car radios. Business and industrial equipment displays represent another significant series of applications. These include analytical instrumentation, automobile clocks and dashboard displays, digital panel meters, global positioning system equipment, and police and various pieces of test equipment.

In the longer range, polymer electroluminescent displays have the potential to become a significant new flat panel display technology for portable and stationary applications. In full-color screens of this type, each color pixel would consist of three individual monochrome subpixels, each of which would emit red, green, and blue. While polymer emitters of each of these colors have been demonstrated in the laboratory, considerable additional development work is required before appropriate groups of different color devices can be integrated in a single display panel. In addition, the development of appropriate driving-addressing technology is still at a very rudimentary stage. To drive daylight viewable displays with VGA-level or higher resolution might require active matrix addressing. A number of research groups in Japan are working on this task.

Polymer electroluminescence also has possible applications in different types of lighting. Today some LEDs achieve, at low voltages, brightness greater than that possible with incandescent or fluorescent lamps. However, at these high brightness levels, the operating lifetime of LEDs is severely limited, partly because of the heat generated by the high power level and a relatively low efficiency. Before some lighting products will be commercially viable, scientists must improve efficiencies by at least an order of magnitude to reduce the generation of heat that occurs with an extremely bright light. A quantum efficiency of 25% or greater is theoretically achievable, and efficiencies in this range could make such large-scale lighting practical.

With further technological advances, conductive polymers have potential applications in optical storage and retrieval devices; products that detect light, such as large-area sensor arrays for military and commercial applications; and high-resolution flat panel displays. The thin-film structure used as an LED also can function as a high-performance photodetector. The asymmetric metal electrodes on either side of the polymer layer provide a built-in potential equal to the difference in their work functions. The photosensitivity of polymer materials can be significantly enhanced by the excited-state charge transfer, using acceptors, such as buckminster-fullerene (C_{60}) or its derivatives.

Highly sensitive polymer photodetectors can be fabricated in large areas by processing conductive polymers from solution at room temperature. They could be made in unusual shapes and on mechanically flexible substrates.

In a somewhat different vein, through a research project sponsored by contracts from the U.S. government, Uniax researchers are developing voltage-controlled resistive networks based on semiconductive polymers. This is a sophisticated application intended to improve the quality of the images captured by an array of visible or infrared light sensors. In the near term, this could impact on advanced sensor systems used by the military for detecting targets in airborne or satellite-based reconnaissance applications.

Another optical application involves using these networks in medical imagery to optimize the image-capturing ability of magnetic resonance imaging (MRI), positron emission tomography (PET), ultrasonic, and X-ray imaging systems.

Another future product application for conductive polymer light emitters is in image-processing boards in a personal computer. Users would have controls, either in software or hardware, that allow them to select different features displayed on the monitor. Such a product would be of interest to individuals involved in analyzing complex large images with many features, such as the satellite imagery of ore deposits. In the longer term, such technology would have application in consumer markets to improve the sensitivity and dynamic range of hand-held video cameras. Polymer LEDs have tremendous long-range potential due to their possible use in flat panel displays.

Another area of commercial interest involves photovoltaic devices. Practical devices have been fabricated using traditional semiconductors, such as cadmium telluride, crystalline silicon, and gallium arsenide. There has been some interest in making photovoltaic devices with crystalline indium phosphide and amorphous silicon. But cells made from amorphous materials do not approach the efficiency of the currently used materials, and experimental cells made with conductive polymers are not as efficient either.

More promising is a future use of conductive matrix resins that would enable aircraft to better tolerate lightning strikes. When lightning strikes an aircraft with a conventional composite skin, the point struck by lightning heats to extreme temperatures. This could cause the surface of the skin to blister or char, or it might even blow a hole through the plane.

Lightning could enter an aircraft at the tail or the trailing edge of its wing. A drop in voltage between the entrance and exit points of the lightning would occur, depending on the aircraft skin's resistance. If the resistance is high, damage to the electronic control system could be severe. Conductive polymers could shield the control system from damage.

On a related front, regions of high conductivity in the skin of an aircraft could contain sensors that measure wing flex, providing a feedback for controlling the airplane's flight. If polymer-based antennas could be imbedded into the skin of an aircraft, they would improve its overall dynamics. If polymer-based power and grounding buses were built into the aircraft's shell, there could be a substantial weight savings gained over similar copper-based devices. (A commercial airliner carries about 1 ton of copper wiring.)

1.9.2 ISSUES AND CHALLENGES

Although there are some conductive polymer-based products on the market and the technology has generated much enthusiasm and optimism among the scientific community for developing others, some impediments still remain that the scientific community must overcome before more products are commercialized. Generally, there is a need to better control structural order in these polymers. In the optical and display area, one of the early issues has involved the ability to produce polymers that have different band gaps. Different gaps would enable the materials to emit red, green, or blue light.

Scientists have addressed this issue. For example, scientists can make poly(p-phenylenevinylene) (PPV) emit blue light by interrupting the conjugation in the polymer with nonconjugated units. They also have been able to create blue LEDs with poly(p-phenylene) as the emitter. To create red light, scientists attach alkoxy side groups to the phenylene rings of the PPV.

But a much more difficult challenge must be overcome before full-color displays using conductive or light-emitting polymers become commercially viable. This entails pixilating the colors. One solution might be to blend different conductive polymers. Each polymer in the blend could have a different band gap. The larger the band gap, the greater the voltage would be needed to inject a charge into the polymer. In this manner, the color emitted from the blend would depend on the amount of voltage applied. This might be one avenue to take in making a full-color polymer-based display.

An alternative approach to pixilation would be to create a microcavity from white light–emitting diodes. The length of such a cavity would determine the color of the emitted light. Prototypes of white light–emitting microcavity LEDs already have been fabricated, but this technique has its shortcomings. The emission wavelength has an angular dependence, which means that as the angle between the viewer and the axis of light increases, the peak wavelength, and the light emitted, change to blue.

In addition, scientists must further optimize the stability and lifetimes of polymer-based LEDs. LEDs produced from other evaporated organic materials perform better in terms of lifetime and efficiencies. But polymer-based LEDs are improving. Researchers at Uniax have achieved polymer-based LED lifetimes of more than 10,000 hours in an inert environment. Single-layer PPV LEDs have lasted up to 3000 hours, although at moderate brightness. Philips scientists have made LEDs from conductive polymers that have thousands of hours of life.

While these advances are very encouraging, the scientific community still has a way to go before commercially viable lifetimes can be achieved. While 3000 hours appears to be a long time, a continuous readout on a clock radio or kitchen appliance would require a lifetime of 8760 hours to last just one year.

The advances being made in understanding what causes polymer-based LEDs to degrade should also help improve their lifetimes. Scientists have learned that oxygen reacts with PPV and other materials, oxidizing them and causing lower efficiencies.

In general, even though polymers can be made conductive by doping or through other routes, they may not be commercially viable if they cannot be made highly conductive or if they are insoluble or difficult to process. For example, in the area of polymer-based thin film transistors, researchers have obtained high mobilities in poly(thienylenevinylene) (PTV).

However, this material is not easy to process. It is insoluble. PTV films have to be made by spinning a precursor polymer onto a substrate and then thermally converting it to PTV at high temperatures using a little hydrochloric acid. But the acid tends to corrode electrodes, which makes the manufacturing steps difficult.

Other issues concern the ability to connect conductive polymers to nano-sized electrodes. Scientists at the Georgia Institute of Technology have focused their conductive polymer efforts on developing molecular wires, conductive paths of atomic orbitals, which would make possible intramolecular electron or hole transfer. Here, an unsaturated chain of atoms acts as a conduit for electrons.

Although the commercial potential of conductive polymers appears almost limitless, it is important to note that few conductive polymers have yet seen commercialization, and only materials that offer unique and necessary performance capabilities over existing materials will find wide use in commerce. ■

2. APPLICATIONS AND MARKETS

Conductive polymer applications that are of primary interest among developers include light-emitting polymers for use in various types of displays, batteries, nonlinear optical materials (for use in telecommunications), photoconductors and semiconductors, sensors, antistatic dissipation, anticorrosion coatings and films, radio frequency interference shielding, and EMI shielding.

Conductive polymers that offer unique and necessary performance advantages over existing materials will find a greater chance of reaching the marketplace. Polymers under investigation exhibit a wide range of electrical and structural properties that potentially make them amenable to application in many fields. Interest and activity in bringing products to market are high in many areas.

2.1 MARKET OVERVIEW

The majority of opportunities for corporate participants in the conductive polymer business will be in the development and sale of end products incorporating the materials rather than in the sale of the materials themselves. The likeliest exception to this will occur when a manufacturer produces conductive forms of plastics that already enjoy widespread industrial use and where conductivity confers some additional performance advantage. The potential benefits from using conductive polymers may be substantial. Most corporate investment returns are likely to occur over a 10-year period rather than a 2- to 3-year time scale.

The market for conductive polymers is still an emerging one. While some applications have reached the commercialization stage, many others remain under development with a goal of commercialization. Some in the field have conjectured that sales of the materials could reach the billion dollar range in the next decade. This could occur if more devices containing these polymers reach market.

One large market would consist of flat panel displays. According to Stanford Resources (Palo Alto, CA), a market research firm, 48.7 million flat panel displays were sold in 1995, which at an average price of $183 per display amounts to $8.9 billion in total sales. This total is projected to increase by an average of 12% annually to 106.3 million units, accounting for $20.8 billion in sales in 2002. Although work remains to be done before a full-color emissive display can be produced using this technology, its potential advantages in weight, power consumption, and cost provide a compelling economic incentive for manufacturers involved in the flat panel business.

The market for anticorrosion applications also stands to be significant, given the need for rust-proofing and maintaining buildings, roads, highways, bridges, and other structures. Sales of conductive polymers to this segment could reach between $50 and $100 million annually, as the technology matures and is more widely used.

On another front, a large market is developing for conductive polymer-based textiles and fabrics. The military could incorporate conductive polymers into camouflage nets to shield armor and vehicles from radar. This market alone is estimated at $4 million.

2.2 APPLICATIONS

2.2.1 LIGHT-EMITTING POLYMERS

Light-emitting polymers used in polymer LEDs promise a radically new approach to the manufacture of a wide range of flat displays. This emerging technology will lead to lightweight and low-power–consuming addressable arrays of light sources for a broad range of applications ranging from simple alphanumeric displays to high-information-content flat panel displays. Conductive polymer technology offers a novel path to the fabrication of thin, flexible displays made entirely of plastic materials.

Considerable attention has been focused on flat panel display technology because of the critical role these devices will play in the consumer electronics and military defense systems of the next decade. Helmet-mounted displays, currently under development for use by combat vehicle crews, are among the important military applications that would benefit from the matrix-addressed polymer electroluminescent displays. In addition, other military programs envision the development of advanced devices that offer automatic target recognition and situation awareness. Each of these devices will require low-power, lightweight displays for the soldier in combat.

The commercial importance of flat panel displays is apparent following the explosive growth of color flat panel displays, made possible by the development of a new commercially viable technology by a number of Japanese companies. This technology, the active matrix liquid crystal display (AMLCD), was first introduced to the marketplace in the mid-1980s.

Continued massive investment in AMLCDs by Japanese and Korean manufacturers in the last decade has given these companies a considerable lead in a business that is very profitable. Nevertheless, undesirable features of these LCDs, including excessive power consumption by backlighted displays and persistently high cost, have led to a vigorous search for alternative display technologies. Among the possible replacements at various stages of development are plasma displays, inorganic electroluminescent panels, and field emission displays.

Until a satisfactory alternative to the AMLCD can be developed, the huge market opportunities in flat panel displays will continue to motivate efforts to find a leap-frog technology. Here, too, Japanese corporate and university laboratories have been active. In October 1995, Pioneer Electronics announced development of alphanumeric displays based on an organic electroluminescent material. The firm has demonstrated some prototype devices.

Other companies have participated in the high-risk–high-payoff search for a leap-frog display technology through development of polymer LEDs. These consist of a layer of conductive electroluminescent polymer sandwiched between a pair of metal electrodes. When a modest voltage is applied between the electrodes, charge carriers are injected into the polymer layer. Oppositely charged carriers meet in the layer and recombine, resulting in the emission of light.

Uniax has made significant progress in developing polymer LEDs by fabricating them on a substrate that exhibits low oxygen and water permeability. The display structure is bonded to a second sheet of impermeable material using a low permeability adhesive. Researchers are considering a number of techniques for forming the bonding seal. Organic polymers like the electroluminescent active layer of these devices must be processed at relatively modest temperatures (about 150 °C or lower). In addition, it is expected that the cost of the bonding method will be an important issue.

All of these considerations suggest that a simple epoxy-based system is an attractive candidate for establishing the bond between the two halves of the proposed display. This type of material is routinely used to form an inexpensive air-tight seal between the two halves of a conventional twisted nematic liquid crystal display cell, although the curing process used in that case involves elevated temperatures that would damage a polymer electroluminescent device.

However, it is known that scientists at the five Japanese companies most aggressively pursuing organic electroluminescent display technology—Pioneer Electronics, Idemitsu/Kosan Ltd., TDK, Sony, and Sanyo—have focused on epoxies to seal their precommercial prototypes. Others have had success in fabricating environmentally robust device packages using a class of filled epoxies that appear to perform well in accelerated age testing.

Scientists at Cambridge Display Technology have examined the operating stability of LEP diodes based on poly(p-phenylene vinylene). They modified a version of PPV to improve its performance in organic electroluminescent devices. Encapsulated single-layer devices with this PPV on a ITO substrate and a calcium cathode have operated in air for more than 7000 hours at 20 °C and for more than 1100 hours at 80 °C without any noticeable degradation. Devices fabricated with this PPV, using a conductive polymer layer as anode and a sputtered low work-function alloy as cathode, operate efficiently and with good brightness at 4 V. They had operating lifetimes of more than 1400 hours.

Considerable attention has been paid to LEDs that utilize thin layers of organic materials as the electroluminescent medium. PPV was the first conjugated polymer shown to exhibit electroluminescence. It has good temperature stability and is intractable. In the absence of chemically reactive species, such as oxygen, PPV is thermally stable up to its decomposition temperature, which is greater than 400 °C.

The availability of a (polyelectrolyte) precursor route makes it possible to solution-process the material at high solution viscosities at a low-solid content. Scientists use such techniques as spin- or blade-coating of the precursor polymer to obtain thin pinhole free PPV films. LEP devices are made by coating the precursor onto substrates, such as ITO-coated glass. This step is followed by thermally converting the precursor to PPV. The polymer layer thicknesses in such LEP devices are of the order of 100 nm.

Cambridge Display has developed a PPV polyelectrolyte precursor that consists of a random copolymer with acetate side groups; tetrahydrothiophenium groups with bromide counterions; and a water/methanol mixture as solvent. The copolymer is deposited on glass and encapsulated with aluminum oxide. Ionic impurities, which cause undesired drift effects in the device or initiate photochemical reactions, are removed to levels below 10 ppm by the dialysis of the precursor polymer.

This process also removes low-molecular-weight compounds. Transition metals and other impurities, which could lead to photochemical instability, are avoided by the using a synthetic processing route.

The LEP devices must be encapsulated. To do this, researchers use a glass slide that is glued over the top of the device with an epoxy resin. Using this PPV copolymer and the glass-epoxy encapsulation, it is possible to fabricate single-layer devices on glass-ITO substrates with evaporated copper cathodes. These have operated reliably for more than 7000 hours (continuously driven in DC current, in air, at room temperature) without noticeable degradation in device performance. Although such devices do not run at high brightness and efficiency, they do operate at high current density.

Cambridge Display's preferred cathode systems are DC magnetron-sputtered alloys with aluminum as the matrix element alloyed with small amounts of low-work-function elements to provide efficient electron injection. It appears that the compact morphology of PPV allows sputter deposition without hurting the device structure. The advantages of sputter deposition include good adhesion to the substrate, high-throughput deposition of pinhole-free films, and compact small grain morphology. It also is easy to deposit alloys of well-defined composition. In addition to the use of sputtered alloy cathodes, a thin layer of a conductive polymer is deposited (spin- or blade-coated) between the ITO layer and the PPV copolymer to improve the efficiency of the device.

Cambridge Display scientists have shown that LEP devices based on a PPV conductive polymer have continuous operating lifetimes of more than 7000 hours at room temperature. Devices with a conductive polymer layer between the ITO layer and the emissive PPV show efficient emission at high brightness, low drive voltage, and good operating stability.

2.2.2 CELLULAR PHONE DISPLAYS

Utilizing simple processing techniques and LEPs, it is possible to fabricate cellular phone displays at less cost and using as little as one-third the power of conventional LED displays. Companies should be able to incorporate a packaged display into a module with drive electronics for substantially less than today's cost of an intelligent display. Moreover, less power consumption translates directly into increased talking time for the user.

An LCD display is typically the most expensive, bulkiest, heaviest, and most power-hungry component in a cellular telephone. Most models available today display between 8 and 20 alphanumeric characters in one to four lines, and may also include icons to indicate the status of functions, such as the state of the unit's battery charge or signal strength.

Today, 80% of cellular telephones use LCDs for their displays. A major disadvantage of LCDs is that they need to be illuminated by a backlight for viewing in low light conditions. This requires a great deal of power. The remaining (20%) share of the cellular telephone display market is occupied by intelligent displays based on matrix-addressable arrays of inorganic LEDs. These consume even more power than nonbacklighted LCD modules, but they can be read in low-light conditions, since they are emissive, and consumers find them attractive.

LED display modules face a major challenge in the near future. The clash between demand for increased information density and demand for lower cost is expected to price them out of the market in the next 5 years. This situation provides an extremely lucrative opportunity for makers of polymer LEDs.

As previously discussed, conductive polymer-based displays are ideally suited to a number of consumer electronic applications, including alphanumeric displays in portable information communication and management devices, such as pagers, inventory control devices, personal digital assistants, and computers. Other possibilities include information read-outs on household appliances, televisions, home and automotive audio equipment, telephones, radios, and clocks, as well as operator interfaces for digital panel meters, vehicular dashboards, analytical instrumentation, and test equipment.

2.2.3 THIN-FILM TRANSISTORS

Thin-film transistors are another application under development for conductive polymers. Among the more likely markets for these plastic transistors are smart cards. Here, reel-to-reel processing of printed electrodes on flexible polymeric substrates should be more economically advantageous than using silicon technology.

The concept of conductive polymer-based transistors appeared in the 1980s. In thin-film transistors, the active semiconductor layer consists of organic or polymeric materials. But only recently have researchers realized that polymeric substrates will not completely replace silicon substrates in transistors, especially since silicon devices are faster. So it appears that any polymer-based transistor will have to find a niche application in which speed is not of utmost importance.

These transistors will find use not in high-speed electronics or computing, but possibly in smart cards or in the bar coding on grocery packages where a flexible plastic substrate would be beneficial. Other potential applications include low-cost, large-area flexible displays and low-end data storage devices.

Researchers at Lucent Technologies and elsewhere have been involved in developing plastic transistors. High mobilities have been reported on a number of conductive polymers, which could be used in transistors. Some of these are not easily processable, such as poly(thienylenevinylene). But others, such as poly(3-hexylthiophene), are. To take complete advantage of the economics of fabricating polymeric transistors, investigators are leaning toward processing them using either spin coating, casting, or printing techniques.

2.2.4 OPTICAL SWITCHING

One way around the electronic data trap in optical systems may have been found by researchers at the University of Utah and Japan's Osaka University, who have developed a polymer-based optical switch that can move data at fast optical speeds without sidetracking them to slow electronic switches. In present optical systems, electronic switches pulse the light on and off. In the new system, beams of laser light trip and reset the switch.

To close the switch, a laser fills the polymer with excitons, the evanescent-charge pairs that block an information-carrying infrared beam. To open the switch, a second laser collapses the pairs, opening the flow again. The process takes just a picosecond. Current optical switches apply an electric field to an inorganic crystal, changing its optical properties to turn a light on and off. They generate light pulses at around 20 GHz. New polymers being designed would boost switching speed within this concept to better than 100 GHz. The new polymer switch could work an order of magnitude faster, at 1 THz.

The basis of the new switch consists of derivatives of PPV that conduct electricity and emit light. The material has been used to make the first polymer-based laser, which absorbs laser light of one color and remits it as a beam of a different color. In the new optical switch, to make the conductive polymer opaque for its off position, it must be hit with a pulse of green laser light. This excites electrons in the material to a higher energy state, leaving behind positively charged electron holes. The newly created energetic electrons and holes hold together to form excitons that absorb infrared wavelength light.

To turn the switch on, the polymer is hit with a pulse from a red laser precisely tuned to stimulate the excitons' electrons and holes into recombining. This makes the polymer transparent to the infrared data beam. In an early demonstration, the researchers only perceived 80 million pulses per second. Getting the control laser up to speed will be required for THz operation, but the switch itself can handle it.

2.2.5 EMI SHIELDING

With the growing demand for computers, cellular phones, and other advanced electronic equipment has come increasing interest by manufacturers in shielding their products from EMI, especially in the radio- and microwave-frequency ranges. In the past, standard metals and composites have been used for shielding equipment from such interference. But these materials have limited physical flexibility, are not lightweight, may corrode, have difficulty in tuning the shielding efficiency, and cannot be easily recycled in today's environmentally friendly atmosphere.

Conductive polymers are good candidates for shielding equipment from electromagnetic radiation or for reducing and even eliminating EMI. The polymers are extremely conductive and have high dielectric constants. These characteristics can be optimized through chemistry. Moreover, they are lightweight compared with conventional materials, are more flexible (being a plastic), and do not corrode as metals do.

Scientists have been able to tailor the microwave frequency conductivity and the dielectric constant of polyaniline by applying different dopants to the polymer, by altering the molecular weight of the molecules and using different solvents for processing the material.

The total shielding efficiency of conductive polymers increases with the thickness of the polymer film. Blends of both conductive and nonconductive polymers also might be incorporated for EMI shielding use.

2.2.6 ANTICORROSION COATINGS

The corrosion of metals, in particular steel, has long been a concern among engineers, city planners, and commercial builders. Over the years, a variety of approaches has been devised for combatting this problem, including treating the surface of steel with chromium, using pickling inhibitors, and applying layers of zinc to the substrate material.

Almost all of these anticorrosion methods function by (1) donating a charge to the substrate metal, such as steel, which is called cathodic protection; (2) withdrawing a charge from the substrate, which is called anodic protection; or (3) sealing off the surface metal from corrosive attacks through the application of a layer of another material.

In addition to being able to protect metals from corroding, conductive polymers represent a solution to the search for an environmentally friendly material that can do the job. Among the polymers studied more extensively for this application is polyaniline. Polyaniline has corrosion-protecting properties in both its doped and neutral states.

The use of conductive polymers, notably polyaniline, for protecting metals is not as simple as it appears. The polymer is not simply coated onto a metallic surface. Complex chemistry is involved in this application. The scientific community first noted that polyaniline coatings offered corrosion protection to metals in the mid-1980s. In the early 1990s, researchers finally developed a more complete knowledge of the process.

Polyaniline behaves like a noble metal because its redox potential is close to that of silver. In this way, polyaniline ennobles the surface of conventional metals, transforming the surface into a thin but dense metal oxide layer. It passivates metals. A polyaniline corrosion coating system includes an ennobling primer and a suitable sealing top coat.

Ormecon Chemie GmbH & Co. KG (Ahrensburg, Germany), a subsidiary of Zipperling Kessler & Co., has introduced a polyaniline that almost completely prevents the corrosion of iron, steel, aluminum, zinc, or other unnoble metals, because it changes their surface properties to make them more noble. If the product, called Corrpassiv, is coated onto metal, the metal oxidizes but does not rust. Instead, an iron oxide layer forms between the metal's surface and the organic metal primer, which acts as a passivation layer, which inhibits the formation of rust.

The Corrpassiv primer contains polyphenylenamine in an extremely fine particle size. Ormecon researchers succeeded in dispersing this neither soluble nor moldable material, so that very fine and even distribution results in the primer, which allows it to serve as a protective primary coating for any metallic part, such as on a boat or ship. A second primer and a top coat also are part of the new system.

Small scratches or other injuries to metallic surfaces are self-healing, thanks to the polymer's ennobling effect. The coating system has been tested successfully in the laboratory and on the open sea. It does not contain poisonous materials, heavy metals, or chromate. The coating can be used for a very thin (20 μm) layer.

The market for anticorrosion coatings is not limited to structures, bridges and the like. Ormecon has developed other anticorrosion coatings under the Corrpassiv trade name for protecting steel and aluminum, for coating screws or other small parts, and for covering machinery components.

Other investigators have found that emeraldine-based polyaniline coatings offered corrosion protection for cold-rolled steel and iron. The amount of protection varied depending on the thickness of the iron oxide layer that developed at the polymer/metal interface and the thickness of the top oxide layer. Researchers achieved the best results when both the top and interfacing oxide layers were removed before the polymer was deposited on the metal's surface. Researchers found that the anodic mechanism of protection occurred. The polyaniline film withdrew a charge from the metal, pacifying its surface against any possible corrosion.

2.2.7 WELDING PLASTICS

Conductive polymers, notably polyaniline, also will find use in welding or joining thermosets and thermoplastics. In practice, either a pure conductive polymer film or a gasket made from a molded blend of the conductive polymer and a thermoplastic or thermoset powder is placed at the interface between the two plastic pieces that are being joined. Microwave energy is applied to the joint, which eventually heats and is fused together.

Investigators have found that the resulting joint may be as strong as the pure compression-molded thermoplastic or thermoset itself. Researchers at Ohio State University (Columbus, OH) have tested this approach when joining two high-density polyethylene (HDPE) bars, using a gasket of HDPE and polyaniline doped with hydrochloric acid. They found that the HDPE bar, placed under tension, yields at a point other than the conductive polymer–formed joint.

Depending on the chemical composition of the conductive polymer and the dopants used, the joint could be permanent or reversible. If the conductive polymer is made nonconductive during the procedure, it will become a permanent joint. If the material remains conductive during the process, the joint will become reversible. When reversing the bond, the conductive material absorbs enough electromagnetic radiation to reheat the joint, making it possible to separate the pieces that had been joined.

2.2.8 CONDUCTIVE TEXTILES AND FABRIC

Scientists have identified a range of applications in which the physical properties of a textile's substrate, such as its strength and flexibility, can be combined with the electrical and microwave properties of a conductive polymer's coating, such as polypyrrole. The resulting polymer-coated textiles could be used as EMI shielding, in composite structures, and in various military applications, such as aircraft skins and radar-protective garments and suits.

Applications for EMI shielding have been developed for a number of areas in which EMI is a major issue, such as in electronics packaging, electronics instrumentation, and power generation equipment.

Textiles have found wide use in reinforced composites. Polypyrrole-coated quartz fabrics used in composite structures with epoxy resins for aircraft wings are similar in strength to noncoated composites. By encapsulating the conductive fabric into a polymer matrix, researchers have been able to make the textile more stable in air.

The microwave response of polypyrrole-coated fabrics make them suitable for such military applications as camouflage, decoys, and aircraft wing edges. The conductive polymer alters the radar signature of the fabric or composite, impeding its detection. Composites made of polypyrrole-coated fabrics behave as a continuous media, unlike granular media with disperse polymer particles. The dielectric behavior of the coated fabrics is better that of a granular composite.

The microwave response to conductive materials depends on their surface resistance. Highly conductive metals are highly reflective. Scientists' ability to tailor the sheet resistance of polypyrrole-coated fabrics allows them to produce materials that have tunable reflection, transmission, and absorption properties. Milliken has developed multispectral fabrics that combine the radar properties of polypyrrole-coated fabrics with visible, near-infrared, and thermal camouflage properties. They use the fabrics as camouflage nets to keep radar from detecting tanks, armor, and other military equipment.

In the production and processing of films and during their use, static charging causes problems. But dispersions of conductive polymers can be used to head off this problem. Using these dispersions, scientists have made films that are permanently antistatic. The polymers form an electrically conductive coating that ensures that a coated surface remains antistatic even under very dry conditions.

The dispersion can be applied in-line using standard coating techniques. Even stretching the film once the coating has been applied is not a problem. The antistatic coating scarcely reduces the transparency of the film. Antistatic properties of conductive polymers might best be applied to video tape, photographic film, and overhead transparencies.

The polyanilines continue to be among the more widely used conductive polymers. A class of water-soluble polyanilines has been developed by oxidatively polymerizing aniline monomers on a template, such as a polymeric acid. These polyanilines dissolve in water. Researchers found that they can apply these materials as removable discharge layers for use in electron beam lithography. After exposure, the polyaniline is removed. Researchers incorporated crosslinkable functionality on the backbone of the polyaniline to create water-soluble polyanilines that can be cured using radiation.

When they are irradiated, the materials crosslink and become insoluble. This makes them ideal as permanent conductive coatings for electrostatic discharge applications as well. Moreover, these crosslinkable polymers could find use as water-developable conductive resists. The polyaniline could be spin-applied onto resists without encountering problems at the interface of the resist. No contamination problems have been experienced during this process either. The water-soluble polymers are an alternative to using organic solvents. The materials can be applied during lithography in a conductive form without the need for any secondary doping steps.

Polyaniline also can be used in the plating through-hole process in the manufacture of printed circuit boards. Printed circuit boards can be single- or double-sided. Holes that are plated through often are a feature of double-sided boards to facilitate the interconnection between the surfaces of the boards and other circuitry.

Copper plating the boards involves laminating copper foil clads on both sides of an insulating substrate. A pattern of holes is drilled into the laminate and then cleaned. A seed, or catalyst—usually a noble metal salt—is applied to the board. The seeded board is immersed into an electroless plating bath in order to plate a thin layer of copper in the plated-through holes. The thin layer of copper renders the plated holes conductive so that a thicker layer of copper can be added later.

However, this process has its drawbacks. The electroless bath is unstable, requiring close monitoring and control. Electroless baths fluctuate between being too stable, which causes voids in the plated-through holes, and being too active, which causes the bath itself to gradually decompose. In addition, formaldehyde, often used as a reducing agent in the baths, is toxic and difficult to dispose. Also, precious metal seeds are expensive and have a limited useful life.

One alternative would be to use a conductive polymer as a conductive electrode for the electrolytic metallization of the copper. Scientists have tried polyaniline in this application. They coated the material to the walls of the holes. A circuit board was immersed in an acetic acid solution of polyaniline for up to 2 seconds. It was dried in a convection oven. The coating appeared thin. Still, the polymer was conductive enough to allow the copper to be deposited.

The polyaniline-coated circuit board was electrolytically plated by immersing it into a copper ion solution. Copper started to plate on the wall of the hole from two sides and grew inward until both sides met at the center of the wall. As the plating continued, a thicker copper coat developed. The time required for the plating procedure depended on the density of the current applied to the system. High-current densities usually resulted in faster plating speeds. The polarization on the circuit board caused a thicker copper coating to develop on the board's surface than on the walls of the holes.

In one instance, IBM scientists plated copper on a board with 45 holes each of which was 1 mm in diameter. They found that polyaniline became almost completely reduced (almost nonconductive) and was not expected to be conductive at the potential when copper is plated. So selecting the right bath for plating is important. The common pyrophosphate copper bath cannot be used for the polyaniline process because the pH of this type of bath usually is too high.

Polyaniline will not become a conductor in this alkaline solution because the polymer will become undoped. It appears that acidic copper baths are more appropriate for this process. In this instance, at the potential at which copper starts to deposit, the polyaniline still is sufficiently conductive. It is not completely reduced to a pH low enough (as low as 6) to allow metallization to occur.

It also is possible to deposit copper on an epoxy substrate that has been coated with polyaniline. IBM investigators first rinsed an epoxy substrate with acetone and then rinsed it with distilled water. Then they blew the substrate dry in a nitrogen stream. An emeraldine-based polyaniline was dissolved in 80% acetic acid and painted onto the substrate at a temperature of about 110 °C to a thickness of about 2 μm.

Using this treatment, the emeraldine base of polyaniline was converted to a conductive emeraldine salt. The coating became conductive. The epoxy substrate with this polyaniline coating was transferred to an electrolyte solution that contained saturated copper sulfate. Using a piece of copper as a counterelectrode, a cathodic current was applied to the polyaniline. An alligator clip in contact with the solution was coupled to the polyaniline-coated substrate. A bright copper film was deposited at the contact point between the clip and the polyaniline coating immediately when the current was applied.

The area of copper film continuously grew and spread outward as the cathodic current was applied. Eventually the entire area of polyaniline was covered with copper. Researchers continued applying the current to increase the thickness of the deposited film. The pH of this copper bath was in the range of 3 to 4. The plating process was not successful in high-acidic copper baths.

The spreading of copper occurred at a faster rate when graphite particles were dispersed in the polyaniline solution and coated onto the epoxy substrate. But it definitely appears that the success and extent of the copper plating depends on how conductive the polyaniline is. Adding graphite does make the process work faster.

2.2.11 CONDUCTIVE ADHESIVES

Conductive polymers have potential use as interconnect materials in electronic devices due to the increased environmental concern about lead contamination and the drive toward more fatigue-resistant materials by electronics device makers. Conductive adhesives have the potential to greatly surpass the reliability of conventional tin lead solders. They can be tailored to meet any requirements that a manufacturer may have. Scientists have improved the adhesives to help resolve significant problems associated with mechanical and electrical properties, such as strength, conductivity, and fatigue life. On the basis of these advances, the electronics industry is looking to use these materials for bonding various components.

Manufacturing electronic surface mount assemblies with conductive adhesives has not been as extensively studied as using solder. Still, some manufacturers of electronic components are considering making a transition from lead-based solders to conductive adhesives. Conductive adhesives' many applications is fueling this move. Potential uses include flexible interconnect attachments on consumer products and die attachment in advanced military systems.

Scientists at the Electronics Manufacturing Productivity Facility (Indianapolis, IN) investigated using conductive adhesives to make a 25-mil pitch surface mount assembly. They found that an adhesive-based process has fewer steps than a lead-based process with most of the time savings occurring in the preassembly steps. The amount of adhesive required for 25-mil pitch applications is less than the amount of conventional solder needed.

In solder paste, the metal by volume is approximately 50%. The remaining paste is composed of volatiles and flux solids that do not serve any purpose in the final interconnect. In adhesives, there are no volatiles or flux to remove when forming the final interconnect. Thus, the amount of adhesive paste required for each assembly is about half that needed when using lead-based solders.

Adhesives can be formulated to have different viscosities. In one project, the viscosity of the adhesive was about 25% that of standard solder paste. Researchers printed the adhesive paste in a manner similar to how lead-based solders are printed. To counterbalance the lower viscosity of the adhesive paste, scientists used a faster printing speed.

A significant difference between the adhesive and the lead-based pastes was in the clean-up. The standard method used to clean up lead-based solder paste involves using isopropyl alcohol and a cloth. The alcohol did not remove the adhesive. Researchers recommended using acetone as the cleaning agent for the adhesive, although extensive exposure of the screen to the acetone caused the screen emulsion to degrade. The adhesive joints created were not ideal. Future efforts will involve optimizing adhesive volume and joint configuration. This specific work centered around developing adhesive assemblies using existing equipment.

Where the components are placed is similar for both solder and adhesive pastes. When using adhesives, manufacturers will have the option of spot (temporarily) curing the materials after components are placed. This would alleviate any component movement during the time between its placement and the curing process. The spot cure has the potential to increase yield and reduce manufacturing costs.

Using adhesives eliminates the need for a solder reflow step. Solders generally require greater temperatures for reflow (about 180 °C) than adhesives need to cure. Researchers cured adhesives at 150 °C for 4 minutes. Manufacturers have the option of using a batch-curing system, but product through-put would be limited by the capacity of the curing system. With adhesives, postassembly cleaning of the mount assembly is unnecessary if proper cleaning procedures are followed during the assembly process.

When using adhesives, there are two processes that require inspection: the actual application of the adhesive paste and the placement of the component. Conventional inspection techniques, including checking for misalignments, are still required when working with conductive adhesives. But postassembly inspection is vastly different than the procedure used for solder.

That which constitutes an acceptable solder joint might not be the same with adhesives. For example, adhesives generally are not shiny, and joint volume is not as important as joint surface area. Manufacturers undertaking full-scale production runs probably would choose to use automated inspection techniques rather than the manual visual inspection performed for this project. However, automated inspection would probably require modifications in inspection equipment because of the different characteristics of adhesives when compared with solder. New inspection algorithms might be required.

2.2.12 CHEMICAL SENSORS

Among the important considerations when developing new sensors is their selectivity and the transduction mechanism by which a chemical interaction is transformed into a measurable physical signal. Several polymers have been used in chemical sensing. Several ion-selective electrodes use polymer matrices that are host specific binding sites, called ionophores or ion exchangers. There is a selective partitioning of ions between the solution and the polymer phase, which is considered to be electroactive because ions have a finite mobility. But the conductivity in these materials often is low.

A new horizon for conductive polymers in sensor applications entails using these materials as chemically selective layers in the sensors. The characteristics that make the polymers attractive for other conductive applications are sometimes less important than those required for chemical sensors. On the other hand, the ability to modify some physical parameters through an interaction with a chemical species is a significant attribute for sensing. This might be considered a secondary doping process, while the insertion of a constituent into the sensing layer during its formation could be considered the primary dopant. The primary doping process, in which an ion is inserted into the sensing layer, occurs in all conductive polymers. It has a major effect on their morphological, mechanical, and permeation properties.

There are three ways that scientists have achieved chemical selectivity in conductive polymers. Selectivity can be derived from the polymer's backbone. Additional selectivity can be obtained by chemically substituting monomers. For example, the polymerization of N-substituted pyrrole results in a material that has less affinity for methanol than unsubstituted polypyrrole. Selectivity also can be achieved through copolymerization of different polymers and blends. One example involves the copolymerization of pyrrole and nitrotoluene. Yet another example entails copolymerizing metalloporphyrine in polypyrrole.

Also, selectivity can be created by preparing the conductive polymer with different doping ions that can be introduced into it after polymerization. Or they exist in the forming solution.

What lends conductive polymers to sensing applications is the ability to prepare a variety of the materials by using similar techniques. This ability facilitates their development. Most polymers can be prepared electrochemically. This approach gives researchers the ability to closely control the deposition of the materials. Close control over the final thickness of the polymer, the ability to deposit the material in a small and geometrically complex area, and the ability to deposit several layers of material at once all complement the trend of sensors toward miniaturization and sensing several chemicals simultaneously.

Researchers at the Georgia Institute of Technology have investigated using conductive polymers like polyaniline in potentiometric sensors. Researchers believe that electrochemically depositing conductive polymers ensures good adhesion of the polymer film to the substrate. The structure and morphology of the polymer coating determines the permeability of absorbent coatings and the diffusion rate of the analyte in the polymer. Conductive polymers will find more sensor applications as their structure and design are more completely investigated.

2.2.13 ELECTRONIC NOSES

Of all the human senses, the ability to smell an odor or aroma has always been the most arbitrary to define. Understanding how the sense of smell functions and analyzing its impact on one's life and environment has long been the goal of researchers. Breakthroughs in sensor technology, coupled with the development of artificial intelligence, has made it possible for scientists to begin to achieve their goals.

Advances in organic chemistry, electronics, and computing have made the measurement and characterization of aromas possible. Systems are capable of producing a digital fingerprint of a smell. The unique qualities and characteristics of many products and compounds are rooted in the chemical volatiles that comprise their odor. The ability to reliably measure and identify impurities, taints, and adulteration is important in many processes and industries.

Traditionally, this difficult task has been the main prerogative of sensory panels whose individual assessment is subjective at best. Analytical techniques, such as gas chromatography and mass chromatography, are sometimes used. But the data gathered are often difficult to correlate with sensory information.

In recent years, electronic, or artificial, noses have been employed to detect and analyze aromas and volatile chemicals in products and the environment. These systems are among the emerging sensor-based uses for conductive polymers. Their main uses lie in environmental and medical diagnostics and the sensory analysis of food products. Electronic noses function by using arrays of sensors that selectively determine specific compounds in an environment. Conductive polymers, such as polyaniline and polypyrrole, find use in artificial noses because they are more sensitive to certain compounds.

Because the sense of smell is important to a physician, an electronic nose could be used as a diagnostic tool in medical applications. An artificial nose identifies problems by examining odors emanating from the body, such as breath or body fluids, or from wounds in the skin. Breath odor can be an indication of gastrointestinal, sinus, or liver problems, as well as infections and diabetes. Infected wounds and tissues emit distinctive odors that can be detected by an electronic nose. Odors from blood and urine may indicate problems with the liver or bladder.

A major market for electronic noses lies among food manufacturers and processors. In some instances, artificial noses augment or replace panels of human sensory experts. Noses can also be utilized to reduce the amount of analytical chemistry required by providing qualitative data. Artificial noses can be used to inspect and grade food quality; inspect fish; control fermentations; check mayonnaise for rancidity; automate flavor control; monitor cheese ripening; verify if orange juice is natural; control microwave cooking; and grade whiskey.

Quality assurance and production control departments have relied heavily on sensory and analytical assessments of odors for many years. Using polymer-based sensors, solid, liquid, and gaseous samples can be objectively characterized in a few minutes. Some systems produce a visual digital fingerprint of an aroma or compound. In quality-control procedures, the technology is used to complement sensory panels. In process control, it detects contamination and minimizes product waste by monitoring aromas throughout the production process. The systems protect valuable brand identity through patenting, a safeguard which up until now has been impractical given the subjectivity of traditional odor evaluation.

Electronic noses are designed to analyze, recognize, and identify volatile chemicals at the parts per billion (ppb) level. Their sensing technology is based on the adsorption and desorption (passing through) of volatile chemicals onto an array of conductive polymer sensors. As with the biological nose, the technology is particularly sensitive to polar species. Amines are most easily detected at the low ppb level.

Monomers are manufactured by an automated process under strictly controlled conditions to produce polymers of uniform spatial geometry. It takes only a few seconds for each sensor to react to a volatile chemical and come to a point of equilibrium (a steady state between adsorption and desorption of the volatiles in the sample). Each polymer in the sensor array exhibits specific changes in electrical resistance, measurable across each sensor element, upon exposure to different odors and aromas. The change in electrical resistance is measured against a predetermined zero reference baseline.

Individual chemical species interact with several sensors. The detection process of each sensor element is nonspecific. The adsorption of a volatile is a result of the polarity (charge) as well as the spatial geometry (size and shape) of the volatile. The better the fit, the greater the electrical change detected. Sensor arrays contain dozens of polymers with different void geometries. This creates a range of overlapping selectivities across the sensor array. One constituent may interact with certain individual sensors, but not with others. This selective interaction produces a pattern of resistance changes, which is charted as a fingerprint.

Concentration also plays a role. One species, present at an elevated concentration, can create a signal equal to that of a low concentration of another species. When an odor is comprised of multiple chemical species, the fingerprint is the sum of their combined interactions with all sensors in the array. The standard array is designed to allow for the detection of a broad range of chemical species.

The interactive data from the sensor array are processed with software that presents the information as a distinctive pattern of responses that can be used as a characterization, or fingerprint, of the odor or aroma. An electronic nose stores unknown aroma patterns and, employing an expert system with artificial intelligence to recognize patterns, identifies or qualifies the odor or aroma.

2.2.14 BATTERIES

Among the more obvious and early applications for conductive polymers are batteries. Conductive polymers that can be switched between an insulating neutral state and a conductive doped state are of interest for these charge-storage applications. A practical charge-storage device using conductive polymers for the anode and cathode requires materials that can be doped and have a high cell voltage. Electrode materials also should be able to undergo several doping and undoping cycles and still retain high charge capacity, good chemical stability, and be very efficient.

Some research groups have developed all-plastic batteries, using conductive polymers in place of traditional electrodes. These batteries are rechargeable and withstand extreme temperatures, making them ideal candidates for space flight and military use. Someday they may be suitable for use in hearing aids and wristwatches.

The objective in this area of development has been to design lightweight batteries that are moldable into just about any shape or size. Researchers also wanted to develop a battery that operates efficiently in extreme heat or cold. Scientists at Johns Hopkins University developed an all-polymer battery in which both of the electrodes and the electrolyte are polymers. The all-polymer battery does not contain heavy metals, which contaminate water and soil. Nor do they contain liquids that pose safety hazards if the batteries leak.

The Johns Hopkins battery has a thin sandwich design that makes the unit easily adaptable. The anode and cathode are made of thin, foil-like plastic sheets. The electrolyte is a polymer gel film that is placed between the electrodes. The cell can be as thin as a credit card.

The Johns Hopkins team substituted functional groups onto the conjugated polymer backbone to achieve molecular-level control of the material's structure and properties. The scientists synthesized a series of fluorophenyl-substituted thiophene monomers to obtain a polymer with high charge capacity and good electrochemical stability in doped and neutral states.

The electrodes for all-polymer batteries were made by electrochemically polymerizing the materials. The electrodes were electrochemically neutralized. Then researchers deposited the polymers onto a thin film of Teflon, which led to the creation of thin, flexible electrodes that did not have any metallic components.

The electrolyte used in the battery was an ionically conductive polymer gel based on polyacrylonitrile. The ability to solution-cast the gel prior to gelation onto the electrodes ensured a good contact between the polymer electrodes and the electrolyte film.

2.2.15 AEROSPACE APPLICATIONS

An attempt to save weight by using lighter materials is driving potential applications for conductive polymers in aircraft and space flight. Electrically conductive composite materials have a variety of aerospace applications. They provide EMI shielding for sensitive electronic controls. Researchers are considering EMI shielding as more microprocessors take over the control functions on aircraft.

Conductive polymers also possess the ability to alter color when voltage is applied to them, which is termed electrochromism. For example, polythiophene changes from a deep blue to a red when oxidized, and back again when the electrical potential is reversed. It only takes a few milliseconds to switch the material from one oxidation state to another. These electrochromic polymers could find use in lightweight, compact video displays, tinted windshields, and visors.

Conductive polymers could be used with aluminized Mylar sheets that are often used as thermal insulation blankets. To ground these blankets, a thin layer of ITO glass is applied to the surface of the Mylar. But applying this coating is a fairly expensive process. Replacing the Mylar with a polymer film that has an electrical conductivity comparable to that of the glass would result in cost savings to manufacturers.

Conductive polymers also have potential as solid reflective surfaces for rf antennas used in communications satellites. Current satellites utilize antenna reflectors coated with vapor-deposited aluminum, which is an expensive process. Conductive polymers, used as the matrix resin in a carbon fiber composite antenna or as a coating on the antenna surface, could be a less expensive substitute.

On another front, highly conductive polymers might be used as the matrix resin for composite rf waveguides. These are currently made from copper or aluminum, or from composites that have an interior coating of pure copper. Copper-coated, carbon fiber epoxy waveguides have been made by copper-plating a precision solid aluminum rectangular mandrel. The composite prepeg is wrapped in the mandrel and cured. Then the aluminum mandrel is dissolved with a caustic solution without damaging the interior copper surface.

This fabrication process would be streamlined if polymer composites with enough conductivity could be fabricated, or if a composite waveguide could be coated with a highly conductive polymer. Such an application would require the polymer to be almost as conductive as copper, but any gains in this area would be significant because manufacturing costs would decline, and there would be some weight savings as well.

Space photovoltaics is another potential application for conductive polymers. Current photovoltaic devices use conventional semiconductor materials. But polymers could be used for applications other than as the main semiconductor. It may be possible to bond a transparent, highly conductive polymer to the front and back surfaces of a semiconductor to form electrical contacts. This may improve the efficiency of the cell because present technology uses metal contacts that block sunlight from a portion of the semiconductor. Even if the material used were not highly conductive, it could be used as a protective cover much lighter than the glass covers now used on photovoltaic devices.

The energy storage systems of photovoltaic devices might be optimized by using conductive polymers. The energy storage portion of the device is usually the most massive. Making this part of the system more lightweight would increase the system's power capacity.

Other applications in photovoltaic devices would harness the optical properties of conductive polymers. The polymers have the ability to frequency-shift light. Because many photovoltaic devices are spectrum-sensitive, it may be possible to increase their efficiency by frequency-shifting infrared light back into the visible spectrum so that it can be more effectively used by the photovoltaic cell. But frequency-shifting efficiencies would have to be optimized for these materials to be useful.

Conductive polymers also can be doped to form n- and p-type semiconductors. As such, they might be transformed into solar cells. This type of solar cell might be less expensive than a conventional unit. The technology already exists that would make it possible to rapidly coextrude these polymers at low cost, but researchers would have to improve the stability of these cells for them to be commercially viable.

The materials also hold potential for use in the electric power industry. The polymers could be used as a surface coating to dissipate a charge over the insulator surface of high-voltage bushings, thus reducing the possibility of flash-over, a major cause of electrical failure. Conductive polymers also could be used as a bulk resin, blended with epoxy, in making bushings that control the internal electric field distribution in equipment and systems.

2.2.16 MOLECULAR CONDUCTORS

In many segments of materials science, researchers are addressing the properties of materials at the molecular level. Electroactive materials, such as conductive polymers, hold much promise. Toward this end, scientists at the Georgia Institute of Technology have investigated making molecular wires, in which an unsaturated chain of atoms acts as a conduit for electrons. Given recent advances in organic synthesis, patterning, and imaging, the molecular-level engineering of solid-state devices based on electroactive linear molecules appears possible.

Researchers have investigated the principles of electrical conductivity in organic polymers, developing a connection between bulk material properties and the properties of discrete organic molecules. They have demonstrated how applying theoretical, synthetic and mechanical organic techniques provide paradigms for studies of partially ordered or semicrystalline materials and insights into the basic conductivity in these materials.

Georgia scientists have examined the properties of discrete molecules as they expand toward a polymeric state. They have concentrated their efforts on polyacetylene because it can be made conductive by a variety of doping methods. Their work provides the basic understanding of how this material functions in a conductive state at the molecular level.

2.2.17 NANOCOMPOSITES

Dissimilar materials, each with ordinary properties, are often weaved by nature into composite materials that have novel characteristics. For example, brittle calcium phosphate and jellylike collagen combine to form lightweight, rugged bone. Chemists would be able to prepare novel, specially tailored materials with unique characteristics if they could synthesize materials at nature's level of sophistication.

Some researchers have combined the self-assembling characteristics of liquid crystals and polymer chemistry to create highly ordered materials that have nanometer-scale architecture. They have made polymer-based composites by developing a self-organizing liquid crystal system, composed of liquid crystal monomers in an organic phase and a conductive polymer precursor in an aqueous phase.

The monomers and precursor have been polymerized into a fixed structure that is similar to an ordered array of conduits filled with wires. The conduits form from a polyacrylate network, and the wires are composed of PPV.

Thermotropic liquid crystals are found in calculator and watch displays. Their degree of order and fluidity varies with temperature. The materials resemble an inverse micelle. The ionic head groups gather around droplets of aqueous solution, and the hydrocarbon tails point outward. The molecules self-assemble, forming an inverse hexagonal array of hydrophilic tubes about 4 nm in diameter.

Each is filled with a solution of PPV precursor. Researchers then heat the polymer, converting the PPV precursor to PPV. The end result is a polymer-based nanocomposite that contains hexagonally ordered tubes threaded with strands of PPV. The material exhibits enhanced photoluminescence. An important step toward an electronic application for the materials will be to see if they electroluminesce.

2.2.18 MICROROBOTICS

A challenge to the development of effective artificial organs and microrobotic components is the creation of machines that can sense, think, and move. Conductive polymers offer a range of properties that make them potentially useful for use in biomedical research, including the fabrication of microsurgical robotics and artificial organ components. Scientists are investigating conductive polymers for use in valves and welds in artificial hearts.

With a significant amount of effort focused on improving the processability and stability of the polymers, researchers have developed polyaniline thin films that are durable. The films can be used as electrodes for growing additional conductive polymer films using electrochemical techniques. While these films have been used to make chemical sensors and electronic devices, ultimately it may be possible to integrate these devices as subsystems in microrobots.

Until recently, polymers have been used only as passive layers in micromachining, serving as patterning, insulating, or mechanically flexible layers. But the emergence of stable, processable conductive polymers brings an opportunity to utilize these materials as active components. Conductive polymers have a variety of interesting properties that can be exploited in microrobotics. These include their ability to change their volume based on their state of oxidation, alter their conductivity, emit light, and store ions. Investigators are developing microactuators and valves on the micrometer scale.

Others have made plastic electromechanical mechanisms. Two polymers with different conductivities change their linear dimensions when current flows through them, similar to the way that metallic strips in thermostats change under different temperatures. The polymers experience more dramatic changes in size, using much less electricity than conventional piezoelectric or electrostatic actuators. Several microactuators coupled together might function as an artificial muscle.

Meanwhile, working to create polymer-based piezoresistive sensors, other researchers are trying to make bridges, virtually strings, of ion-implanted conductive polymer films. This gives them the ability to make miniature vacuum gauges, infrared sensors, and strain gauges. A bridge thermally isolates the film, showing large resistance changes with temperature changes and increasing the sensitivity of the devices.

A precursor polymer, poly(styrene-co-acrylonitrile), is dissolved, filtered, and spin-coated onto insulating substrates to form the films. Conductivity is induced in the polymer by ion implantation with 50 KeV nitrogen ions. High-quality films with resistivities from 400Ω/square to 10 MΩ/square can be produced. Suspended conductive polymer bridges are formed by etching a sacrificial layer underlying the film. The bridges are about 200 times as long as they are wide and are about 1500Å thick. The bridges are durable enough to withstand etching, high-speed rinsing, and high-pressure air drying. The films themselves have already shown excellent resistance to solvents, water, acids, and bases.

Applications of suspended bridge structures include Pirani gauges for vacuum measurement, flow meters, anemometers, thermal conductivity analysis, and gas detection. They all require the ability to monitor changes in the dissipation constant of the device, and in each case it is good to thermally isolate the resistor from its substrate. ■

3. RESEARCH EFFORTS AND OPPORTUNITIES

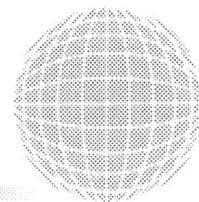

The organizations and individuals listed and reviewed here were contacted and interviewed for this report—they are involved in the cutting-edge of research in conductive polymers. Contact them if you have questions about their work, or if you are interested in collaborating with them or in licensing their technology to facilitate your own product development.

Researchers at the University of California, Berkeley, have been developing new routes for synthesizing conductive polymers. In addition, they have developed polymer composites with nanoscale control, which has resulted in a molecular wire. Their technique leads to highly ordered materials that have nanometer-scale architecture.

The chemists combined a self-organizing liquid-crystal system and a conductive polymer precursor and then polymerized the compounds. They obtained a fixed structure resembling an ordered array of conduits filled with wires. The conduits are a polyacrylate network, and the wires they contain are the conductive polymer PPV. This polymer has been used to fabricate polymer-base light-emitting diodes and lasing materials.

The water-deficient system resembles an inverse micelle. The material is not crystalline, but flows with the component molecules exchanging places. Photopolymerizing the hydrophilic channels locks the architecture in place. Enhanced photoluminescence also is present.

Commercial Opportunities: Researchers are applying for patents covering the molecular wire work. The technology may have optical applications. Scientists are seeking to commercialize their technology.

Contact: Douglas Gin, University of California at Berkeley, Department of Chemistry, 202 Lewis Hall, Berkeley, CA 94720. Phone: 510-642-7756. Fax: 510-643-1846. Internet: gin@chemgate.cchem.berkeley.edu.

Scientists at the University of Florida have investigated the synthesis, electrochemical behavior and electrochromic properties, of a series of conju-

3.1 UNIVERSITIES AND RESEARCH ORGANIZATIONS

3.1.1 UNIVERSITY OF CALIFORNIA, BERKLEY

3.1.2 UNIVERSITY OF FLORIDA

gated conductive polymers, including alkyl-derivatized and arylene-based polymers. They also have developed several dual-polymer electrochromic devices based on new low– and high–band gap conductive polymers.

For such electrochromic applications as windshields and mirrors, researchers want to alter the redox state and absorbance properties of conductive polymers, which would enable the materials to change color. It would then be possible to develop windows that change from clear to opaque in order to reduce the sun's glare, or rearview mirrors that lessen the intensity and glare of headlights at night.

Additionally, investigators have examined the long-term switching redox of polypyrrole. One of the most important criteria for many polypyrrole applications is its stability during prolonged redox switching. This especially comes into consideration when the polymer is a candidate for a charge-storage material in batteries and electrochemical capacitors or as a switchable film in electrochromic devices.

Investigators found that ion-solvent exchange processes and long-term stabilities of polypyrrole are affected by the supporting electrolyte and the solvent used for electropolymerization and redox switching. The film composition determines the permeability of dopant ions when different solvents are used during electropolymerization and redox switching.

The long-term electroactivity during redox switching of polypyrrole is determined in part by solvent size or polarity differences between polymerization and redox switching, as well as ionic conductivity. The best long-term switching ability was obtained when the same electrolyte and solvent were used during both electropolymerization and redox switching.

Commercial Opportunities: Scientists will consider collaborations with industry to commercialize their work.

Contact: J. R. Reynolds, Department of Chemistry, Center for Macromolecular Science and Engineering, University of Florida, PO Box 117200, Gainesville, FL 32611. Phone: 352-392-9151. Fax: 352-392-9741. Internet: reynolds@chem.ufl.edu.

3.1.3 GEORGIA INSTITUTE OF TECHNOLOGY

Research into conductive polymers is moving ahead at Georgia Institute of Technology. A group of scientists is investigating the use of conductive polymers as molecular wires in molecular electronics. While they are continuing their work on developing processable polymers, their research also involves examining conductive phenomena at the nanometer scale.

One Georgia team is attempting to determine what conductivity at the molecular level means. Researchers are trying to learn if oligomers with built-in charge carriers are superior to ones that must be externally doped. Recent experimental evidence from different laboratories indicates that connecting gold surfaces with an unsaturated chain does not lead to "electronic communication" between the gold surfaces. Researchers believe this may change if the unsaturated chain contains a charge carrier, also known as a soliton. A challenge they face entails connecting conductive polymers to nanoelectrodes.

The scientists are looking into the production of films and fibers, as well as coatings on fibers used in textiles. Prototypes still must be produced.

Another group of Georgia Tech scientists is aiming to integrate conjugated conductive polymers with other electronic and photonic materials, as well as fabricate micro- and nano-scale features of these materials for use in developing electronic devices. In order to prepare materials with optimized properties, they are investigating the use of molecular self-assembly techniques to organize small arrays of molecules.

Although conductive polymers, such as polyacetylene and polyarylene, should find use in sensors, batteries, and EMI shielding, they have some drawbacks. The materials are characterized by a low mobility of charge carriers, slow redox switching, and microstructural defects. Some of these problems have been overcome by substituting their rigid polymer backbones with flexible side chains to create soluble materials that have well-defined microstructural homogeneity.

Others at the university are examining sensor applications for conductive polymers. Some of their effort has entailed demonstrating the versatility of thin layers of polypyrrole, polyaniline, and other conductive polymers in potentiometric chemical sensors for detecting chemical gases and vapors. A typical application will involve monitoring analytes in the workplace.

Researchers have developed chemical sensors that they can demonstrate in the laboratory and possibly in the field as well. Investigators may be able to have a commercially viable prototype within a year, unless major problems with the technology arise. They have been working with polypyrrole, polyaniline, and polyphenylene, among other materials.

Among the challenges they face is to develop a better understanding of how the materials function and the capacitance phenomena. They also would like to better determine the principles that govern the thickness of the material's layers and the doping-undoping process.

Commercial Opportunities: Investigators will consider establishing collaborations with companies to commercialize their efforts.

Contacts: On molecular wires: Laren Tolbert, Chairman, School of Chemistry and Biochemistry, Georgia Institute of Technology, Atlanta, GA 30332. Phone: 404-894-8222. Fax: 404-894-7452. Internet: laren.tolbert@chemistry.gatech.edu.

On conjugated polymers: David Collard, Associate Professor, Georgia Institute of Technology, School of Chemistry and Biochemistry, Atlanta, GA 30332. Phone: 404-894-4026. Fax: 404-894-7452. Internet: david.collard@chemistry.gatech.edu.

On sensors: Mira Josowicz, Principal Research Scientist, Georgia Institute of Technology, School of Chemistry and Biochemistry, Atlanta, GA 30332. Phone: 404-894-4032. Fax: 404-894-7452. Internet: mira.josowicz@ chemistry.gatech.edu.

3.1.4 JOHNS HOPKINS UNIVERSITY

Researchers at Johns Hopkins University have developed an all-plastic battery, using polymers in place of the conventional electrode materials. The battery, which is rechargeable and environmentally friendly, has military and space applications and may soon be suitable for small consumer devices, such as hearing aids and wristwatches. *Popular Science* magazine honored the battery with a "Best of What's New" award, naming it one of the top 100 new products, technology developments, and scientific achievements of 1996.

The project was initiated and funded by Rome Laboratory (Rome, NY), a U.S. Air Force research and development center. Air Force officials asked Hopkins to create lightweight plastic batteries that could be molded into almost any size and shape for use in satellites and military equipment. Scientists produced polymers that can generate up to 2.5 V in cells that potentially could compete with the 3 V lithium batteries now on the market.

As part of this collaborative effort, engineers at Hopkins' Applied Physics Laboratory have been paving the way for practical systems applications by linking the batteries with a solar cell charging system. Fabricating an all-plastic cell was difficult because most polymers that conduct electricity lack sufficient energy difference to serve as electrodes.

Batteries consist of three main components: an anode (the positive electrode), a cathode (the negative one), and an electrolyte (the conductive material between the electrodes, such as the liquid in a car battery). Although other researchers have used polymers for one of these components, Hopkins scientists are among the first to create a practical battery in which both of the electrodes and the electrolyte are made of polymers.

Lab tests indicate that the cells can be recharged and reused hundreds of times without degradation. The all-plastic battery operates efficiently in extreme heat or cold. In addition, the power cell's thin sandwich design makes it highly adaptable to different shapes and sizes. The cell can be as thin as a business card. This could allow battery users to cut a cell to fit a specific space. The Hopkins group has applied for patents.

Commercial Opportunities: The developers are interested in finding nonmilitary applications for their battery technology. It is available for licensing.

Contacts: Theodore Poehler, Vice Provost for Research, Johns Hopkins University, 276 Garland Hall, 3400 N. Charles St., Baltimore, MD 21218. Phone: 410-516-8765. Fax: 410-516-8035. Internet: top@jhu.edu. D. Dylis, USAF Rome Laboratory, 32 Hangar Rd., Rome, NY 13441. Phone: 315-330-4587. Fax: 315-330-1531. Internet: dylisd@rl.af.mil.

Investigators at the NASA Lewis Research Center have optimized a technique for generating composite films of both conductive and nonconductive polymers. The films are composed of a conductive polymer absorbed into the surface of polyimide films. Pyrrole and 3-methylthiophene were evaluated as precursors for the conductive phase of the material.

Researchers derived empirical models for each precursor to describe the effects of different polymerization variables on the conductivity of the films.

Commercial Opportunities: Researchers are open to corporate collaborations.

Contact: Mary Ann Meador, Environmental Durability Branch, Materials Division, NASA Lewis Research Center, MS 49-3, 21,000 Brookpark Rd., Cleveland, OH 44135. Phone: 216-433-3221. Fax: 216-977-7132. Internet: maryann.meador@lerc.nasa.gov.

3.1.5 NASA LEWIS RESEARCH CENTER

Researchers at Ohio State University have been involved in the conductive polymer field since its infancy. Their efforts have been spearheaded by Arthur J. Epstein, professor of physics and chemistry and director of the Ohio State University Center for Materials Research. Epstein's group carries out both fundamental and applied research in metallic, semiconductive, and magnetic polymers. The nature of the group's basic research involves basic charge transport and optical and magnetic studies. Epstein and colleagues often study applications in parallel with their fundamental studies.

For example, they examine conductive polymers for electrostatic dissipation and EMI shielding applications in parallel with studies of conductivity mechanisms. They have investigated LEP devices and polymers for anticorrosion coatings while simultaneously studying electronic phenomena in semiconductive polymers. The research team studied polymers for

3.1.6 OHIO STATE UNIVERSITY

use in magnetic shielding and induction applications while performing fundamental studies on magnetism in the polymers.

While Epstein and colleagues have made prototypes for each of these applications, light-emission and magnetic shielding applications are still a few years away from commercialization.

Specific projects undertaken by Epstein and colleagues include investigating the shielding efficiency of various conductive polymers as a function of their conductivity and dielectric constant; controlling the dielectric response of polyanilines; examining the corrosion-preventing capability of polyaniline; studying emissions in bilayer polymer LEDs; examining color-variable bipolar light-emitting devices; studying the electronic control of pH at sulfonated polyaniline electrodes; and reviewing the insulator-metal transition and the inhomogeneous metallic state in conductive polymers.

Meanwhile, other scientists at Ohio State have proven the feasibility of a large-scale process for producing a conductive polymer in an inexpensive and environmentally friendly manner. They doped a polyaniline powder with camphorsulfonic acid (CSA) using supercritical carbon dioxide as a carrier gas. The process was successfully completed, but not as easily as they had predicted.

CSA did not dissolve in supercritical carbon dioxide alone, but required a mixture of water and ethanol. Then the investigators attempted to measure the solubility of the acid in the mixture, but their results varied considerably. They attempted to impregnate the polyaniline with CSA by flowing the mixture through the polymer. The amount of polyaniline decreased during the run due to loss of polyaniline from the polymer vessel. The experiment was repeated at a reduced flow rate and increased pressure to prevent polyaniline from being transported out of the polymer vessel.

The experiment produced positive results, and investigators are using the data to generate a model of diffusion in a sphere. This model will allow them to optimize the process to achieve a specific concentration of the CSA in the polymer during full-scale production.

Executing several runs at different flow rates showed that the solubility of CSA in pure carbon dioxide was inadequate. The polymer vessel was modified by using a 7 millipore filter instead of a paper filter to hold the polyaniline. The first run with the paper filter caused polyaniline to be blown through the filter and collect on various components downstream from the polymer vessel.

Containing the polymer in the 7 millipore filter prevented most of the material from being carried through the system, but the volumetric flow rate had to be dropped further to completely eliminate loss of polyaniline.

Researchers compensated for lowering the volumetric flow rate by increasing the pressure from 1800-2800 psi so that the carbon dioxide could dissolve a higher density of CSA.

The experimental work can be divided into two main areas: determining the CSA solubility in the supercritical carbon dioxide and the primary doping of the polyaniline with the CSA. Five trials were performed to determine the solubility of the CSA in supercritical carbon dioxide. Of the first four runs, three actually resulted in a negative weight change. The difficulties and errors in these trials were caused by the apparent low solubility of the CSA. After these trials, investigators decided to use HPLC grade water and ethanol as cosolvents to help facilitate the process.

The fifth run resulted in positive results. The CSA precipitate was observed as a white dust on the tip of the needle in the needle valve. Under different circumstances, more trials would be needed to verify an accurate CSA solubility. However, due to the scope and time constraints of this project, this last successful trial was used to determine the solubility of CSA in supercritical carbon dioxide, which was 7.99 mg CSA per 200 mL of carbon dioxide.

After calculating the CSA solubility in the supercritical carbon dioxide, the next step entailed attempting a primary doping of the polyaniline sample with CSA. The sixth overall run (the first doping run) gave very unexpected results. Upon completion, researchers found that a large quantity of the polymer had escaped from the contacting vessel and had been deposited throughout the remainder of the system, with the bulk ending up in the inlet to the needle valve that follows the outlet of the system.

For the seventh run, the polymer was placed inside a 7 millipore filter within the filter housing to try to improve its containment. While this run provided good results, the eighth run demonstrated a negative change in the polymer weight. This was due again to some loss of polymer from the contact vessel.

In preparing the dopant solution, researchers expected the solubility of the CSA in supercritical carbon dioxide to be adequate, because the increase in density of supercritical carbon dioxide increased its solvent characteristics. CSA was much more soluble in a mixture of carbon dioxide, water, and ethanol. With this solvent system, the desired solubility was achieved. When researchers attempted to dope the polyaniline with the supercritical solution, they expected the weight to increase in the dry sample after the process. The weight of the dried product increased by 0.2265 g over the initial starting material. This indicates that a sufficient amount of CSA adsorbed into the polyaniline particles.

For application to process design, adding cosolvents to the supercritical fluid gave an insight into commercial-scale equipment requirements and operating conditions. The process begins with a mixing tank, into which the carbon dioxide, water, and ethanol are metered in and mixed. After the liquid solution is prepared, the liquid is pumped to a heat exchanger and heated to supercritical conditions: a temperature of 313 K and pressure of 1800 psi.

Once the fluid is at supercritical conditions, it feeds to a column packed with the CSA. In this column, the supercritical solution becomes saturated with the dopant. Exiting the CSA contactor, the fluid proceeds to the polyaniline contactor where the doping process takes place. The fluid contacts the particulate polyaniline and the CSA diffuses into the polymer particles. A recycle loop was inserted to conserve carbon dioxide. After the doping is complete, the bulk of the carbon dioxide is flashed off, condensed, and fed back into storage. The water, ethanol, and CSA are removed from the remaining solution.

Based on a 5.5 L vessel for the polyaniline contactor, the total equipment cost is approximately $252,000. Using industry-based estimates of installation, control, piping, and electrical costs, the initial capital investment for implementation of the process is about $600,000.

One way to improve the process might be to test different forms of polyaniline in the contactor. Spherical particles are easy to model and are available for experimental and industrial use. Scanning electron microscopy images of the supercritical carbon dioxide–treated polymer show no visible physical changes in the material.

Commercial Opportunities: Scientists will consider collaborating with industry to further develop their process.

Contacts: Arthur J. Epstein, Distinguished University Professor, Professor of Physics, Professor of Chemistry, Director, the Ohio State University Center for Materials Research, Ohio State University, 174 W. 18 Ave., Columbus, OH 43210. Phone: 614-292-1133. Fax: 614-292-3706. Internet: epstein.2@osu.edu.

David L. Tomasko, Assistant Professor, Chemical Engineering Department, Ohio State University, 121 Koffolt Laboratory, 140 W. 19 Ave., Columbus, OH 43210. Phone: 614-292-4249. Fax: 614-292-3769. Internet: tomasko@er6.eng.ohio-state.edu.

3.1.7 UNIVERSITY OF MICHIGAN

Researchers at the University of Michigan have investigated *N*-methylated poly(nonylbithiazole), a dopable, conjugated poly(ionomer). They also have patented a process for producing thiazole polymers. The methylated polymers can be reversibly doped and have been employed in building LEDs and lithium batteries. Investigators have found that the electrical, optical, and mechanical properties of the materials can be tuned by methylation of certain atoms, resulting in new conjugated poly(ionomers).

The U.S. patent (5,536,808) discusses the chemistry of these polymers, including their cyclic units, and how each of the units is connected to another thiazole unit.

Commercial Opportunities: Scientists are seeking to license their technology.

Contacts: M. David Curtis, Department of Chemistry, University of Michigan, 930 N. University Ave., Ann Arbor, MI 48109. Phone: 313-763-2132. Fax: 313-647-4865. Internet: mdcurtis@umich.edu.

Licensing: Technology Management Office, University of Michigan, 3003 S. State St., Wolverine Tower, Rm. 2071, Ann Arbor, MI 48109. Phone: 313-764-8202. Fax: 313-936-1330. Internet: jsrobts@umich.edu.

3.1.8 UNIVERSITY OF RHODE ISLAND

University of Rhode Island investigators are working on anticorrosion applications for conductive polymers. A passive layer develops between a conductive polymer coating and aluminum alloy surfaces. Aluminum alloys are lightweight, high-strength materials that have a 1% copper impurity, causing them to corrode and lose some mechanical strength.

Investigators have coated aluminum alloys with a double-strand conductive polymer coating of polyaniline and have found that a third interface develops between the polymer coating and the metal. Researchers removed the polymer and found that this interfacial layer has a dense and smooth oxide-type surface morphology. It is conductively stable.

They first synthesized the polymer complex. It was dried, yielding a black powder that is soluble in water or in an alcohol-based solvent. Additional processing changes the polarity of the second strand's functional groups to make the complex soluble in an organic solvent. Investigators applied the soluble material to small centimeter-sized samples of aluminum either by dipping the alloy in the material, by coating it onto the surface, or by dropping the coating onto the metal from a pipette. Coating thicknesses ranged up to about 50 µm.

Researchers believe that the protective film is either a polymer-modified oxide layer or a chelated complex of the polymer film to aluminum ions or aluminum metal surface atoms. Oxygen is present under the layer of polymer film. The energy spectrum of the aluminum is not typical of bohemite or an aluminum oxide. One hypothesis is that the aluminum is chelated with the polymer to form a modified oxide passive layer.

When a metal, such as an aluminum alloy, is exposed to a corrosive environment, the polymer layer self-passivates as a rechargeable battery does and continues to oxidize. The conductive polymer appears to withstand acidic environments. As anticorrosion coatings, conductive polymers also could be used on copper and cold-rolled steel.

On another front, university scientists have investigated using polyaniline as an electrochromic material. This material has a conductive green-colored state that is chemically stable in air and in aqueous solutions. Polyaniline also appears suited for use in electronic displays, showing at least three reproducible changes in color.

Commercial Opportunities: University of Rhode Island scientists are looking for industrial support.

Contact: Sze Yang, University of Rhode Island, Department of Chemistry, Kingston, RI 02881. Phone: 401-874-2377. Fax: 401-874-5072. Internet: syang@chm.uri.edu.

3.1.9 NATIONAL UNIVERSITY OF SINGAPORE

The National University of Singapore is conducting basic research into conductive polymers. The research is undertaken as part of the program at the university's Materials Science Department.

Commercial Opportunities: Researchers are interested in collaboration projects with industry.

Contact: Head, Department of Materials Science, Faculty of Science, National University of Singapore, Singapore 119260. Phone: +65-7722610. Fax: +65-7776126. Internet: masngsc@leonis.nus.sg.

3.1.10 SOUTH BANK UNIVERSITY LONDON

South Bank University London established a Unit for Specialty Electronic Polymers in 1994 to promote applications of specialty electronic, or conductive polymers and to assist industry in exploiting uses for the materials. Researchers are working in different areas, including large-area flat panel displays, electroluminescent polymers, LEPs, transparent conductors, sensors, EMI shielding, conductive silicones, and conductive gaskets.

Scientists at the university, with financial backing from Nissan Ltd., are developing conductive polymer thin films for display applications, such as computer and mobile phone displays. They also are involved in the development of LEPs for flat panel displays. Another effort entails developing electrochromic displays with the materials. One application they are specifically working on is camouflage, in which the color of the polymer can be made to match the surrounding environment.

They also have investigated conductive polymers as microwave absorbers to microwave-weld thermoplastics. The organic nature of the polymers helps to form good bonds between the pieces of thermoplastics that are being welded. The university research team also has produced a capacitor based on niobium pentoxide, using polyaniline and polypyrrole as the cathodes.

In the area of EMI shielding, a technique was developed for directly electrodepositing nickel and polypyrrole onto carbon-impregnated nonwoven polyesters. The process resulted in the continuous formation of tapes that can be used as EMI shielding materials. The tapes have high enough conductivity and shielding efficiency and low enough surface-transfer impedance to be used as cable shielding materials.

Commercial Opportunities: The Unit for Specialty Electronic Polymers, along with the university's Electromagnetic Compatibility Centre, offers consultancy services to industry. Analytical and production facilities also are available.

Contact: P. Kathirgamanathan, Director, Unit for Specialty Electronic Polymers and Electromagnetic Compatibility Centre, South Bank University London, School of Electrical, Electronic and Information Engineering, 103 Borough Rd., London SE1 0AA, UK. Phone: +44-171-815-7545. Fax: +44-171-815-7599. Internet: kathirp@sbu.ac.uk.

3.1.11 UNIVERSITY OF TEXAS AT ARLINGTON

A major thrust of conductive polymer research at the University of Texas at Arlington is focusing on the multistep synthesis and study of transparent conductive polymers with low band gaps, which means that less energy than usual is needed to make them conductive materials. Usually, conductive polymers are colored in their insulating state. Upon conversion by oxidation to the conductive state, they become more colored and opaque, often dark blue-black.

In the university's laboratory, several low–band-gap polymers have been prepared and studied. These have their light absorptions shifted to longer wavelengths, toward the red. When they are doped, the light absorption shifts into the infrared, making the materials much lighter, such as light yellow, and transparent in some cases.

Another research area involves LEPs. Studies are underway in the synthesis of such new polymers. Some investigations include developing ways to improve the lifetimes and efficiencies of electroluminescent LEDs and analyzing new device designs.

Additionally, scientists are looking at new ways to markedly improve the electrical conductivity of these polymers. One approach is to prepare polymers that are structurally similar to known organic conductors and super-conductors. Another approach entails the new synthesis of known electrical conductors so that they are defect-free at the molecular level. They should have much greater conductivity, and with proper processing, such as spinning into fibers, might rival some metals in their electrical conductivity.

Specifically, Texas researchers have developed luminescent polymers that have discrete emitter units. They also have synthesized a new low–band-gap polymer poly(2-decylthieno[3,4-b]thiophene). In addition, they have synthesized and performed electroluminescence studies of a polythiophene with an attached carbonyl group. Researchers fabricated electroluminescence devices by spin-coating the polymer onto ITO-coated glass and then depositing a layer of aluminum. Various thiophenes emitted different colors, including a red-orange light.

Others at the university are investigating anticorrosion coatings. They base their research, as do others, on the ability of conductive polymers to anodically protect metals. They have found that the corrosion-altering effects shown by conductive polymers in electrical contact with such metals as steel are caused by both the polymer and iron being electrochemically active in corrosive environments.

Different types of conductive polymers exhibit varied electrochemical behavior that depends on the chemical composition of the polymer and the applied potential. An electrochemical process common to all of these materials involves the reversible doping-dedoping process that is attributed to reversible oxidation and reduction of the conjugated polymer's backbone. This process takes place at different electrochemical potentials for such structurally different conjugated polymers as polypyrrole and polyphenylene.

Polymers that oxidatively dope at higher potentials have correspondingly greater chemical potentials in their doped state. They are stronger oxidizing agents. When in contact with metals, p-doped polymers provide a galvanic couple whose electrochemical potential is determined by the chemical potential of the doped polymer and the metal with which it is in contact. Doped polymers with high chemical potentials strongly anodically polarize metals. This influences the degree to which a passivating metal forms a passivating oxide layer and then reforms this layer when it is scratched or damaged.

Other factors to be considered include the reversible and irreversible electrochemical processes that occur in a conductive polymer during the corrosion of the metal in the galvanic couple. The extent to which irreversible processes may occur will determine the useful lifetime of the polymer as a corrosion-preventing material.

Commercial Opportunities: Texas researchers are looking for industrial collaboration opportunities.

Contacts: On low-band gap polymers: Martin Pomerantz, Professor, Department of Chemistry and Biochemistry, University of Texas at Arlington, Box 19065, Arlington, TX 76019. Phone: 817-272-3811. Fax: 817-272-3808. Internet: pomerantz@uta.edu.

On anticorrosion materials: Ronald Elsenbaumer, Professor, Department of Chemistry and Biochemistry, University of Texas at Arlington, Box 19065, Arlington, TX 76019. Phone: 817-272-3812. Fax: 817-272-3808. URL: http://www. utachem.uta.edu.

3.1.12 TNO INSTITUTE OF INDUSTRIAL TECHNOLOGY

TNO has established a Division of Materials Technology, where researchers are performing basic research and exploring applications for conductive polymers. These include using the materials in EMI shielding, conductive inks and adhesives, antistatic coatings, actuators, and batteries.

One project is focusing on using conductive polymers for the galvanic deposition of metals on plastics. The project is sponsored by the Netherlands Ministry of Economic Affairs. It involves using polypyrrole and thiophene as primers for depositing metals on nonconductive plastic substrates. These polymers may make it possible to speed the metallization process because they are conductive.

Researchers have studied different steps that are part of the metallization process: pretreating the epoxy glass substrate; depositing the conductive polymer; and depositing the metal. Investigators have studied how the pH of the polymerization solution affects the electrical resistivity of the primer.

Commercial Opportunities: TNO is interested in performing contract research for industry.

Contact: Roland van de Leur, TNO Institute of Industrial Technology, Division of Materials Technology, PO Box 6031, NL-2600 JA Delft, Netherlands. Phone: +31-15-2696558. Fax: +31-15-2696501. Internet: R.vandeLeur@ind.tno.nl. URL: http://www.tno.nl.

3.1.13 UNIVERSITY OF TSUKUBA

Researchers at the University of Tsukuba are involved in basic research of conductive polymers, but they also are interested in applications work. They have been studying synthesis of highly conductive polyacetylene thin films; synthesis of aligned thin films of polyacetylene using liquid crystalline solvents; use of crystalline compounds for synthesizing aligned conjugated polymers; development of catalysts for polymerizing acetylene; and synthesis of ferromagnetic conjugated polymers.

Commercial Opportunities: Researchers are collaborating with industry on synthesizing conductive polymers and are interested in pursuing other collaborations.

Contact: Hideki Shirakawa, Professor of Polymer Chemistry, Institute of Materials Science, University of Tsukuba, Tsukuba, Ibaraki 305 Japan. Phone: +81-298-53-5103. Fax: +81-298-55-7440. Internet: hideki@riko.tsukuba.ac.jp.

3.1.14 University of Utah

Scientists at the University of Utah have developed a conductive polymer–based optical switch that can move data at fast optical speeds all the way, without sidetracking it to slow electronic switches. Beams of laser light trip and reset the switch. To close the switch, a laser fills the polymer with excitons, the evanescent-charge pairs that block an information-carrying infrared beam. To open the switch, a second laser collapses the pairs, opening the flow again. The process takes a picosecond.

The basis of the new switch are derivatives of PPV. The material has been used to make a polymer-based laser, which absorbs laser light of one color and remits it as a beam of a different color. One problem involves the tendency of PPV to break down when hit repeatedly with laser light. The polymer films also heat up when they absorb infrared light, which may degrade them further.

Others at the university are investigating artificial organ and robotic applications for conductive polymers. They have made polyaniline thin films that can be used as electrodes. Ultimately it may be possible to integrate the materials as subsystems in microrobots and artificial organs.

Commercial Opportunities: Researchers will consider industrial collaborations.

Contacts: On optical switching: Sergey Frolov, Professor, Physics Department, University of Utah, Salt Lake City, UT 84112. Phone: 801-581-4402. Fax: 801-581-4801.

On artificial organs and robotics: Douglas Chinn, Department of Materials Science and Engineering, University of Utah, 281 EMRL, Salt Lake City, UT 84112. Phone: 801-581-5676. Fax: 801-581-4816. Internet: douglas.chinn@m.cc.utah.edu.

3.2 Companies

3.2.1 Abtech Scientific Inc.

Scientists at Abtech have been developing sensors based on conductive polymers. In the devices, the polymer serves as an active or field-responsive component. These may be chemical or biological sensors that exploit a large change in electrical conductivity of the polymers upon their reaction with certain dopants. Among the applications are sensor devices used to measure glucose in whole blood and urea in blood dialysate.

These devices link multifunctional polypyrrole copolymer films with the biocatalytic action of the enzymes glucose oxidase and urease. This is achieved through a reaction with hydrogen peroxide and the generation of bicarbonate, ammonium, and hydroxyl ions. Researchers have developed enzyme-linked immunosensors and DNA biosensors by using polypyrrole-immobilized antibodies and oligonucleotide sequences.

Also under development are gas sensors based on the sorption and partitioning of gases and vapors into electroconductive polymers and the alteration of their impedance properties. Such sensors have given rise to an electronic nose for detecting chemicals and measuring volatile organic compounds. The developers have used an array of conductive polymer sensors to measure levels of toluene and other compounds. These chemicals would be of interest in emissions control. Another area of interest has been drug delivery. Researchers have developed an electroactive hydrogel actuator that controls the release of bioactive peptides.

Fundamental work also continues, aimed at a better understanding structure-property relations in the polymers. In addition, scientists are investigating ion and hole-electron transport in thin films of pyrrole, aniline, and thiophene.

Commercial Opportunities: The company is interested in developing strategic alliances with companies that want to commercialize some diagnostic products. Abtech would look favorably on relationships to which Abtech brings its diagnostic technology and a partner brings marketing and sales capabilities. Both parties would invest dollars in product development, which would be followed by a commercialization stage that involves marketing. Suitable exit and separation strategies would be incorporated into all agreements.

Contact: Anthony Guiseppi-Elie, Scientific Director, Abtech Scientific Inc., 1273 Quarry Commons Dr., Yardley, PA 19076. Phone: 215-321-3256. Fax: 215-321-7099. Internet: guiseppi@abtechsci.com. URL: http://www.abtechsci.com.

3.2.2 AMERICHEM INC.

Americhem has a number of conductive polymer products on the market. The products are based on Versicon, a doped polyaniline. The company is marketing transparent conductive coatings and processable compounds that can be used in molding and extrusion. The coatings are liquid dispersions of Versicon in specific film-forming matrices. The coatings are designed to adhere to plastic substrates. They dissipate charges before they accumulate and create an electrostatic shield that prevents charges from reaching electrostatic discharge–sensitive materials.

Commercial Opportunities: Americhem is marketing materials.

Contact: V.G. Kulkarni, Americhem Inc., 723 Commerce Dr., Concord, NC 28025. Phone: 800-331-8412. Fax: 704-784-1034. Internet: vgkulka@aol.com. URL: http://www.americhem.com.

3.2.3 AROMASCAN INC.

AromaScan markets multielement sensor systems that emulate the human nose. They can be used for analyzing aromas and a variety of chemicals. Applications include food and pharmaceutical analysis and environmental monitoring. The systems generally comprise an array of sensors and a gas-sampling device that directs a vapor of uniform concentration to the array.

In many instances, the applications for these electronic, or artificial, noses will determine the type of sensors they will contain. Usually, conductive polymers are selected as the sensor material in devices that will be used to analyze polar species, such as flavors, fragrances, and aromas. The polymers respond better than other materials to these types of compounds.

Commercial Opportunities: The company is marketing electronic noses, some of which utilize conductive polymers.

Contact: Technical Department, AromaScan Inc., 14 Clinton Dr., Hollis, NH 03049. Phone: 603-598-2922. Fax: 603-595-9916. URL: http://www. aromascan.com.

3.2.4 BAYER CORP.

Bayer is marketing antistatic coatings for photographic film and other applications. The coatings are made from 3,4-polyethylenedioxythiophene (PEDT). Also on the market is 3,4-ethylenedioxythiophene, which is the monomer used in making the conductive polymer. In addition, PEDT can be used as an electrode in solid electrolyte capacitors, as well as for the through-hole plating of printed circuit boards.

Company researchers may investigate additional applications for conductive polymers in semiconductor manufacturing and different types of displays.

Commercial Opportunities: PEDT is available commercially.

Contact: Joseph Morrison, Manager, Technical Service and Applications Development, Inorganic Chemicals, Bayer Corp., Industrial Chemicals Division, 100 Bayer Rd., Pittsburgh, PA 15205. Phone: 412-778-4305. Fax: 412-778-4301. URL: http://www.bayer.com.

3.2.5 CAMBRIDGE DISPLAY TECHNOLOGY LTD.

Cambridge Display Technology (CDT) was spun out of the Cavendish Laboratory at the University of Cambridge in 1992. The firm is involved in research and commercial development of LEPs. CDT currently has 25 employees, 20 of whom are dedicated to research. CDT was founded after initial work led by Richard Friend and Andrew Holmes at the university. They discovered that light-emitting structures could be made from polymers as opposed to traditional semiconductor materials.

Scientists found that the polymer poly(p-phenylenevinylene) (PPV) emitted yellow-green light when sandwiched between a pair of electrodes. Initially this proved to be of little practical value as it produced an efficiency of less than 0.01%. But by changing the chemical composition of the polymer and the structure of the device, an efficiency of 5% was achieved, bringing it well into the range of conventional LEDs.

Cambridge University is one of CDT's largest shareholders. This is a pioneering step for the university and, indeed, for academic institutions in the UK, who have been reluctant to invest their own funds in the commercial exploitation of research. CDT has secured a number of other investors from a diverse range of backgrounds, including Cambridge Research and Innovation Ltd; the Genesis rock group; the Generics Group plc; Hermann Hauser, a founding director of Acorn Computer, now an Olivetti company; amd Steve Kahng, president of Power Computing Corp. CDT recruited a chief executive officer, Danny Chapchal, in March 1996 to develop and implement exploitation plans for the company and its technology.

CDT executives announced the firm's commercialization strategy to bring its LEP technology to market in 1996. Since then, the company has signed an agreement with Philips Electronics BV and a technology access agreement with Hoechst AG (Frankfurt, Germany). Under the terms of the Hoechst agreement, in exchange for an up-front fee from Hoechst, CDT will grant licenses on agreed terms to Hoechst's customers.

The agreement covers the precise license terms to be granted to Hoechst's customers and includes royalties on product sales. Hoechst manufactures the polymer fundamental to the construction of an LEP display. The technology licensing agreement means that Hoechst can provide its key customers with the core technology to enable them to begin development of LEP displays and access to key technology held by CDT.

Hoechst had already made a considerable, undisclosed, investment in LEP technology. CDT has partnerships with Uniax and Philips Electronics to develop LEP technology. Uniax is developing small emissive display products based on LEP technology for portable, battery-powered applications.

The first fruit of the Hoechst-Philips-Uniax collaborations is a $2 million clean room and prototype manufacturing line for small, approximately 1 x 2 inch, light-emissive displays, expected to be operational and producing low quantities of displays in the near term.

Philips and Hoechst completed a multimillion dollar minority investment in Uniax in February 1996. In addition to the equity investment, Philips Electronics and Hoechst each has codevelopment contracts with Uniax. Both companies have substantial internal development programs underway. The near-term target is to replace the existing backlights for LCDs in applications where space, voltage, and power consumption are at a premium, such as in mobile telephones. Eventually, displays will be developed for use in consumer products that now depend on LEDs or LCDs, such as personal digital assistants, CD players, electric razors, alarm clocks, radios, and ultimately televisions.

CDT's relationship with Philips will give Philips access to CDT's LEP technology. Philips is paying CDT an upfront license fee together with a royalty on all LEP products. In return, Philips is gaining access to CDT's technology for the lifetime of its patents. The agreement opens the possibility for Philips to scale up existing laboratory processes and to develop its own manufacturing techniques and processes for the manufacture of specific small displays.

Potential benefits of this technology over current technologies (such as LEDs and LCDs) will eventually enable Philips to manufacture thinner displays that operate at lower voltages and consume less energy.

Under terms of an agreement with Uniax, CDT is giving Uniax access to its global LEP intellectual property. In return, Uniax is dropping its dispute of CDT's patents and will pay a royalty fee on all future products.

In the past year, the Cambridge Display research team has synthesized polymers that emit light in the red, green, and blue regions of the visible spectrum. Work is underway to develop driving schemes that allow these materials to be used to construct full-color graphic displays without the need for a complex active array of electronic switches.

Research and development efforts are currently focused on extending device lifetime and reliability; developing more efficient light-emitting structures; designing manufacturable processes; and developing effective drive schemes for graphic displays.

Commercial Opportunities: The company's longer-term objective is to enter different graphics display markets, currently valued at more than $20 billion. CDT's exploitation route for the technology is through licensing and technology transfer, coupled with corporate partnerships. CDT executives realize that the company has not developed the manufacturing and marketing skills essential for competing in the business. Through licensing, the company intends to bring other manufacturers up to speed in LEP technology. This will allow them to develop specific products for their markets.

CDT also is looking for partners to develop LEPs for high–information-content graphics displays and flexible substrates suitable for LEP use. CDT has started to implement this exploitation strategy with the Philips Electronics agreement.

Contact: Mark Gostick, Cambridge Display Technology Ltd., 181a Huntingdon Rd., Cambridge CB3 0DJ, UK. Phone: +44 1223 276351. Fax: +44 1223 276402. Internet: mgostick@cdtltd.co.uk. URL: http://www. cdtltd.co.uk.

3.2.6 DSM BV

DSM is marketing ConQuest water-based conductive coatings based on polypyrrole. The coatings are aimed at antistatic and anticorrosion applications. Since they are water-based materials, they are more environmentally friendly and are less difficult to apply to surfaces.

A ConQuest coating system consists of a ready-to-use dispersion in which the particles of a binder are individually enclosed within a shell of conductive polymer. When this dispersion is applied to an insulating material, it forms a plastic coating that conducts electricity. The coating is transparent, flexible, and not soluble in water. The dispersed particles are small, making it possible to produce coatings less than 1 µm thick. The surface resistivity of the coated substrates depends on the thickness of the coating. ConQuest can be used to make films permanently antistatic, such as in the production of video tapes, photographic film, and overhead transparencies. The material does not allow the films to acquire a static charge during their production or use.

In another application, the ConQuest conductive dispersion is added to corrosion-protection primers. The materials can be applied to passivate steel, aluminum and copper. Even if it is scratched after it is coated, the metal remains protected by the polymer.

Electrostatic coatings work on conductive materials like steel, but nonconductive materials, such as plastics, first need to be rendered antistatic. This can be done by using carbon black–based primers or by modifying the base resin, but in both cases the mechanical properties of the plastic part are compromised. Applying a small coating of ConQuest does not affect the mechanical properties of the plastic substrate. A dry coating thickness of 1 µm or less is sufficient.

The antistatic qualities of ConQuest also make it suitable for materials and articles used in dust-free environments, such as in the clean rooms found in the electronics industry. Additionally, DSM scientists have been developing a coating system that functions as a heat-reflecting layer for use on glass or plastic sheets. Polypyrrole reduces energy transmission in the infrared region. Transparent materials coated with thin layers of ConQuest show a relatively high reflection of infrared light. The coating transmits most of the visible light, so the coated material remains transparent. Infrared-reflective coatings form a barrier to solar heat and for this reason are used in glazing and plastic sheeting.

Commercial Opportunities: DSM is marketing ConQuest on a worldwide basis.

Contact: Marc A. M. M. van Doorn, Business Project Manager, DSM BV, Het Overloon 1, Heerlen, PO Box 6500, 6401 JH Heerlen, Netherlands. Phone: +31-45-5782606. Fax: +31-45-5782266. Internet: m.a.m.m. doorn-van@research.dsm.nl. URL:http://www.dsm.nl.

3.2.7 HUGHES AIRCRAFT CO. RESEARCH LABORATORIES

Scientists at Hughes Aircraft Co. Research Laboratories have investigated polyaniline and polythiophene as the active transducer in environmental sensors. They deposited thin films of polyaniline across gold-interdigitated electrodes. Volatile organic compounds modulated the conductivity of the transducer.

The detection of these compounds is caused by a structural perturbation in the conductive polymer, which is caused either by a direct interaction of the polymer with a pollutant or because of a structural change in the counterion with which the polymer is associated. Such sensors could be fabricated into discrete components of multisensor arrays. The developers are testing a prototype sensor.

Commercial Opportunities: Hughes researchers would consider collaborating with companies on applications for conductive polymers.

Contact: Frederick Yamagishi, Senior Research Scientist, Batteries and Polymeric Sensors Department, Sensors and Materials Laboratory, Hughes Aircraft Co. Research Laboratories, Loc. MA, Building 254, M/S RL-70, 3011 Malibu Canyon Rd., Malibu, CA 90265. Phone: 310-317-5747. Fax: 310-317-5484. Internet: fyamagishi@msmail4.hac.com.

3.2.8 IBM

Scientists at the IBM Thomas J. Watson Research Center have been investigating electronics applications for conductive polymers. They have looked at using polyaniline for in-hole plating, or metallization, of printed circuit boards. They also have found the material suitable for several lithographic applications.

Polyaniline appears to be an effective discharge layer in resist systems for electron-beam lithography and a good removable discharge layer for use in the high-resolution inspection and dimensional measurements of X-ray and optical masks. In addition, the IBM team has used polyaniline anodes to increase the brightness and lifetimes of polymer light-emitting diodes.

Commercial Opportunities: IBM is marketing a product called PanAquas. It is a polyaniline for use in electronics packaging to dissipate electrical charges.

Contact: Marie Angelopoulos, IBM Thomas J. Watson Research Center, PO Box 218, Yorktown Heights, NY 10598. Phone: 914-945-2535. Fax: 914-945-2141. URL:http://www.ibm.com.

Scientists at Lucent have been examining transistor applications for conductive polymers. They have used poly(3-alkylthiophene) and screen printing techniques to fabricate high-mobility plastic transistors. The field-effect mobilities of these transistors are similar to those formed on silicon-based substrates.

3.2.9 LUCENT TECHNOLOGIES

Researchers found that the screen printing method is a simple and environmentally friendly way to fabricate electronic circuitry and make interconnections. It enables patterns to be formed in a single step. The printing process significantly reduces the time and cost associated with photolithography. Using a liquid-phase processable organic semiconductor allows investigators to produce less expensive electronic devices with flexible plastic substrates for displaying or storing data.

While plastic transistors cannot compete with silicon transistors in general terms of performance, the polymer devices could find niche applications, such as in the barcoding of products at stores. The polymeric substrate is flexible and can bend while the silicon substrate would break on a package. Scientists are in the basic research stage of their work. Such an application is at least 5 years away from commercialization.

Commercial Opportunities: Lucent scientists are willing to discuss aspects of their work in the public domain.

Contact: Zhenan Bao, Lucent Technologies, Bell Laboratories, 600 Mountain Ave., Room 1D-246, Murray Hill, NJ 07974. Phone: 908-582-4716. Fax: 908-582-3609. Internet: zbao@bell-labs.com.

Milliken is marketing polypyrrole-coated textiles that can be used for static dissipation, EMI shielding, microwave and radar shielding, and resistive heating, among other applications. For electrostatic dissipation, these Contex conductive textiles can be used in conveyor belts, garments, shoes, wrist bands, flooring, carpeting, pipelines, furniture, aircraft escape slides, high-speed rollers and brushes, and air filters.

3.2.10 MILLIKEN RESEARCH CORP.

The textile absorbs radar waves, altering readouts on radar screens, concealing the location of an item in question. For example, the fabric can be cured into the composite wing of an aircraft, or it can be used in radar-protective garments.

Commercial Opportunities: Contex fabrics are available in widths up to 72 inches and in all lengths up to 300 yards. Milliken is willing to custom-coat fabric within these parameters. Contex yarns are available in 2-lb. packages. Samples are available. The price for these products was not available. The developmental price for Context textiles was $45 per yard.

Contact: Hans Kuhn, Milliken Research Fellow, Milliken Research Corp., M-405, 920 Milliken Rd., PO Box 1927, Spartanburg, SC 29304. Phone: 864-503-2320. Fax: 864-503-2417. Internet: janet_miller@milliken.com.

3.2.11 MONSANTO CO.

Monsanto Co. has acquired from AlliedSignal Inc. the latter's worldwide patent portfolio for conductive polymers. The acquisition is part of a broader Monsanto commercialization program aimed at developing a full line of conductive polymer products with industrial applications.

Monsanto intends to manufacture and supply the Versicon conductive polymer, which had been marketed by AlliedSignal. Versicon is an insoluble powder form of polyaniline used as a conductive additive to polymer blends. Previously, Monsanto developed a process for making a soluble, heat-stable form of polyaniline that can be blended with other polymers without using hazardous solvents.

Among the conductive polymer applications that Monsanto scientists have investigated are electrostatic dissipation, corrosion protection, and sensors. Monsanto scientists have investigated conductive polymers that, when added to anticorrosion coating primers, protect steel, even when the protective coating has pinholes and minor scratches. Such coatings are especially useful in the maintenance industry for coating equipment and piping at chemical processing plants, pulp and paper mills, ships, offshore oil rigs, and automobiles.

Commercial Opportunities: Monsanto is making samples available.

Contact: Robert E. W. Jansson, Vice President, Technology, Monsanto Co., 800 N. Lindbergh Blvd., St. Louis, MO 63167. Phone: 314-694-4804. Fax: 314-694-9058. Internet: rejans@monsanto.com.

3.2.12 NEC CORP.

NEC is marketing a conductive polymer-based capacitor that offers lower impedance and higher frequency, which may make up for the fact that it could cost 20–100% more than conventional tantalum capacitors. It incorporates polypyrrole, which increases the product's conductivity. The material also increases the product's decomposition temperature and improves its reliability. The Neocapacitor has the same structure as a conventional chip tantalum capacitor. It contains a low-resistance cathode of functional polypyrrole that substitutes for the manganese dioxide used in a conventional capacitor.

Commercial Opportunities: The Neocapacitor is available for sale worldwide. It is used in cellular phones, camcorders, and some PCs.

Contact: Fumito Takayanagi, Senior Manager, Solution Engineering, Circuit Components Division, NEC Corp., 7-1, Shiba 5-chome, Minato-ku, Tokyo 108-01, Japan. Phone: +81-3-3798-9628. Fax: +81-3-3798-6152. URL: http://www.nec.co.jp.

3.2.13 NEOTRONICS SCIENTIFIC INC.

Neotronics Scientific markets electronic noses that use polypyrrole-, polyaniline-, and polythiazene-based sensors. Utilizing Neotronics conductive polymer sensors, the aroma or vapor is characterized, then compared to a known pattern, or fingerprint. Variations between sample and fingerprint highlight differences in the vapor from the norm.

Conductive polymers are best used when polar species are the target of detection, such as flavors, aromas, and fragrances. The nose utilizes conductive polymer sensors that have rapid response times and stable outputs. The reaction of the vapor with the conductive polymer causes a change in the material's conductivity. This change is dependent on a complex interaction between the components of the vapor and the polymers, as each sensor responds to a number of components in a unique manner. The use of an expanding range of polymers makes comparative analysis of complex vapor structures possible.

In a typical nose, a 12-sensor array is located in a sealed compartment, where it can be purged with a gas in order to obtain a baseline response and to refresh the sensor head after it is exposed to a sample. The sample is placed in a glass vessel in a temperature control cell beneath the sensor head compartment. The sample can be heated and stirred with an integral magnetic stirrer before and during analysis.

The headspace above the sample is purged briefly to eliminate environmental odors from the atmosphere in which the sample was taken. When the headspace above the sample has equilibrated, which could take up to several minutes, the sensor head lowers out of its sealed compartment into the vapor above the sample. Measurements from the sensors are taken at this point, and the sensor head is raised automatically after the analysis is completed. The system may continue to collect additional replicates on the same sample or can prompt the operator for a new sample.

The fingerprint patterns obtained from the instrumentation are displayed in several graphical formats, such as bar or polar plots. Computerized comparisons of the plots give the user a measure of how different two samples are or how different a sample is from the stored reference pattern.

Commercial Opportunities: Neotronics is marketing several electronic nose systems incorporating conductive polymers.

Contacts: Ley Hathcock, Director of Research and Development, Neotronics Scientific Inc., 4331 Thurmond Tanner Rd., Flowery Branch, GA 30542. Phone: 770-967-2196. Fax: 770-967-1854. Internet: hathcock@ neotronics.com. URL: http://www.neotronics.com.

In Europe: Neotronics Scientific, Western House, 2 Cambridge Rd., Stansted Mountfitchet, Essex, CM24 8BZ UK. Phone: +44-1279-814848. Fax: +44-1279-813926. URL: http://www.neotronics.co.uk.

3.2.14 NESTE OY CHEMICALS

The main efforts at Neste have been directed at creating a processable polyaniline to be used in such melt-processing methods as fiber spinning, film blowing, and injection molding for polyolefin-based polymers. Researchers have tested their material in applications with other companies and have supplied material both for research purposes and preliminary production runs. But the polymer is not yet fully commercialized.

In addition to providing conductivity, the conductive polymer compositions improve melt processing performance by lowering the melt viscosity and the processing temperature and by shortening the processing time. This facilitates the injection molding of complex-shaped objects.

Commercial Opportunities: Neste scientists may consider applications development with other companies.

Contact: Matti Jussila, Manager, Applications Development, Conductive Polymers, Neste Oy Chemicals, Technology Center, PO Box 310, FIN-06101, Porvoo, Finland. Phone: +358-204-50-7031. Fax: +353-204-50-7724. Internet: matti.jussila@cc.neste.com.

3.2.15 ORMECON CHEMIE GMBH & Co. KG

Ormecon is a subsidiary of Zipperling, a one-time German compounder and producer of masterbatches and specialty compounds. Zipperling has sold its compounding business and is exclusively focusing on the research, development, and marketing of organic metals through its Ormecon subsidiary.

Ormecon is marketing a dispersible conductive polyaniline. The company is targeting anticorrosion applications under the Corrpassiv trade name. The product prevents the corrosion of iron, steel, aluminum, zinc, or other unnoble metals by changing their surface properties to make them more noble.

If one coats iron with a Corrpassiv primer, the iron oxidizes, but does not change to rust. Instead, an iron oxide layer is formed between the iron surface and the organic metal primer, which acts as a passivation layer. The formation of rust is inhibited. The primer contains polyphenylenamine, a pure organic powder in an extremely fine particle size. Ormecon scientists succeeded in dispersing this neither soluble nor moldable material, so that a very fine and even distribution results in the primer, enabling it to serve as a protective primary coating for any metallic part, for example on a boat.

A second primer and a top coat are part of the system. Small scratches or other injuries are self-healing, as the ennobling characteristics are effective in up to 2 mm thick coatings. In addition, under the Corrrpassiv trade name, Ormecon scientists have developed several other primer and coating systems for protecting steel and aluminum and for use in dipping and spinning techniques used to coat screws or other small parts.

While most of the company's effort has been aimed at anticorrosion applications, the firm also produces transparent conductive coatings.

Commercial Opportunities: Ormecon researchers will consider collaborative and consultation opportunities with other companies.

Contact: Bernhard Wessling, Ormecon Chemie GmbH & Co. KG, Kornkamp 50, PO Box 1464, D-22904 Ahrensburg, Germany. Phone: +49-4102-4900-18. Fax: +49-4102-4900-52. Internet: wessling@ zipperling.do.eunet.de. URL: http://www.ormecon.de.

3.2.16 PANASONIC INDUSTRIAL CO.

Panasonic is marketing the EEF-CD, a surface-mount, specialty polymer aluminum electrolytic conductor. It is suited for bypassing voltage supply lines in computers and wireless telecommunications products. The ideal situation is to have a stable, unchanging voltage presented to electronic circuitry. The reality is quite different. Bypassing voltage supply lines creates a lower impedance that more effectively stabilizes supply voltage to the circuitry.

Usually, aluminum electrolytic capacitors have a limited lifetime due primarily to a loss of electrolyte. But this component contains a specialty polymer (polypyrrole) dielectric that improves its lifetime.

Commercial Opportunities: Panasonic is marketing these polymer capacitors.

Contact: George Harayda, Marketing Manager, Panasonic Electronic Components, Panasonic Industrial Co., Applied Technologies Group, Panazap: 7E-2, Two Panasonic Way, Seacaucus, NJ 07094. Phone: 201-392-6197. Fax: 201-392-4315. Internet: haraydag@panasonic.com. URL: http:// www.panasonic.com.

3.2.17 PHILIPS COMPONENTS BV

Philips has been a player in the conductive polymer arena both on its own and with others via technology agreements. As previously discussed, Cambridge Display Technology Ltd. and Philips Components signed a nonexclusive agreement giving Philips Components access to Cambridge's LEP technology. Under the terms of the agreement, Philips is paying Cambridge an upfront license fee together with a royalty on all LEP products. In return, Philips is gaining access to Cambridge's technology. The agreement opens the possibility for Philips to scale up existing laboratory processes and to develop its own manufacturing techniques and processes for making specific small LEP displays.

LEP displays are constructed by applying a thin film of the polymer onto a glass or plastic substrate coated with a transparent ITO electrode. An aluminum electrode is deposited on top of the polymer. Applying an electric field between the two electrodes results in emission of light from the polymer. Philips and Hoechst AG also have placed a multimillion dollar minority investment in Uniax Corp., which is profiled later in this chapter. In addition to the equity investment, Philips Electronics and Hoechst each have codevelopment contracts with Uniax.

Also, scientists at Philips Research Laboratories have investigated polymeric microelectronic devices and plastic field-effect transistors. Scientists also have performed field-effect measurements in doped conjugated polymer films and have deposited, using molecular beams, thin films of pentacene for organic field-effect transistor applications.

Commercial Opportunities: Philips researchers may be willing to discuss aspects of their conductive polymer research that are in the public domain.

Contact: I. Martin Schuurmans, Managing Director, Philips Research Laboratories, Prof. Holstlaan 4, 5656 AA Eindhoven, Netherlands. Phone: +31-40-2744135. Fax: +31-40-2743645. URL: http://www.us.philips.com/research/reslabs/natlab.

3.2.18 STRATEGIC DIAGNOSTICS INC.

Strategic Diagnostics has developed an immunosensor based on conductive polymers that screen for the pesticide atrazine. The portable instrument detects the pesticide at the ppb level. Electrodes coated with the conductive polymer are housed in an injection-molded three-well cartridge.

Contact: Strategic Diagnostics Inc., 128 Sandy Dr., Newark, DE 19713. Phone: 302-456-6789. Fax: 302-456-6770. URL: http://www.sdix.com.

3.2.19 UNIAX CORP.

Uniax was founded in 1990 by two professors at the University of California (Santa Barbara, CA). After successful development and licensing of processing technology for conductive polymers in 1994, company researchers focused on developing initial products: LEP displays. Following an equity investment by Philips and Hoechst, Uniax was able to accelerate the development effort, resulting in the early 1997 demonstration of medium-information-content emissive displays. A prototype manufacturing line in Santa Barbara will enable Uniax to fabricate displays of various sizes and configurations.

In early 1997, Uniax completed construction of a clean room and prototype manufacturing line for small (approximately 1 x 2 inches) light-emissive displays. At a cost of $2 million, the prototype line is intended for producing low quantities of displays. These small dot matrix displays consisting of several thousand pixels are targeted for hand-held battery-powered applications. The polymer displays are expected to have significant advantages over backlighted LCD displays and alphanumeric LED displays in cost and performance.

The polymer light-emitting technology is being developed by Uniax in conjunction with its corporate partners, Philips and Hoechst, who completed a multimillion dollar minority investment in Uniax. Philips is developing this technology for a variety of applications for both internal and external use. Hoechst is providing both partners with state-of-the-art polymers for these products and expects to become a major supplier of the conductive polymers and films required by the industry as it develops.

The core technology of Uniax is its scientists' expertise in synthetically altering certain semiconductive or metallic polymers to produce forms of the material that are processable. These can be used to fabricate novel electro-optical devices. Uniax is primarily focused on polymer light-emitting devices (Fig. 3.1).

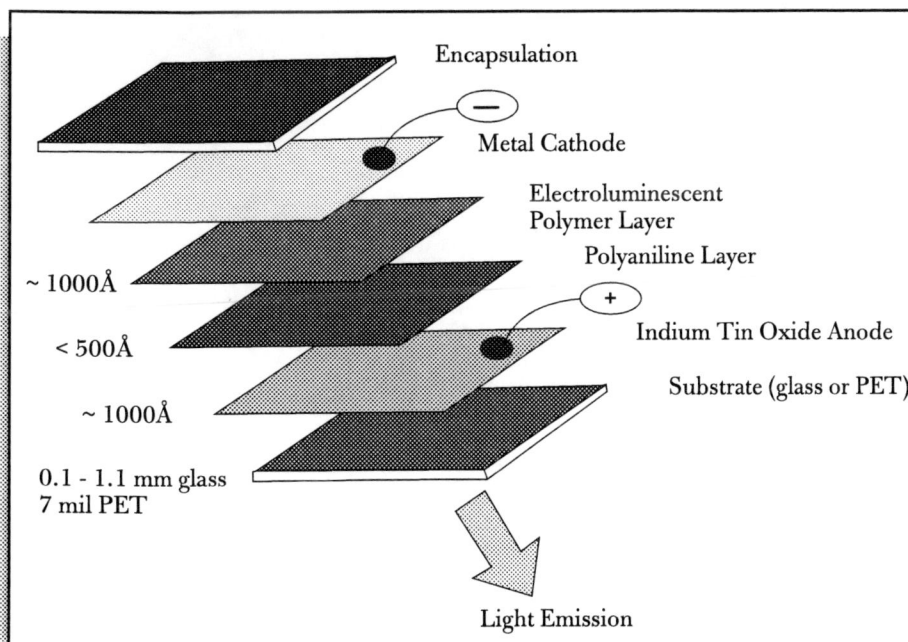

Encapsulation

Metal Cathode

Electroluminescent
Polymer Layer

Polyaniline Layer

~ 1000Å

< 500Å

Indium Tin Oxide Anode

~ 1000Å

Substrate (glass or PET)

0.1 - 1.1 mm glass
7 mil PET

Light Emission

Figure 3.1
Light-emitting device.
Permission of Uniax Corporation.

Commercial Opportunities: Uniax is interested in pursuing contract research and development work. Previous projects have ranged in length from about 6 months to more than 2 years. They have involved the participation of from one to more than five full-time scientists and engineers. Such efforts have included the counterion-induced processing of polyaniline and light-emitting electrochemical cells.

Contact: Jim Long, President, Uniax Corp., 6780 Cortona Dr., Santa Barbara, CA 93117. Phone: 805-562-9293. Fax: 805-562-9144. Internet: info@uniax.com. URL: http://www.uniax.com. ■

APPENDIX A. RECENT U.S. PATENTS

The following patents represent applications of interest to scientists that are developing commercially viable uses for conductive polymers.

5,253,100. Solid electrolytes for conducting polymer-based color switchable windows and electronic display services. Issued: Oct. 12, 1993.

Inventors: Sze Yang and Jyun Hwei. Assigned to Board of Governors for Higher Education, State of Rhode Island and Providence Plantations, Providence, RI. An electrochromic material, polyaniline, is polymerized in situ in a polymeric electrolyte to form an electrochromic/polyelectrolyte mixture. The mixture is coated as a film onto electrochromic material.

5,352,574. Surface-functionalized and derivatized electroactive polymers with immobilized active moieties. Issued: Oct. 4, 1994.

Inventor: Anthony Guiseppi-Elie. New variants of electroactive and optoactive polymers are formed from the surface chemical modification of free-standing and substrate-supported polymer films. The free-standing or substrate-supported films are chemically modified at or near their surfaces to introduce hydrophilic and/or reactive functional groups, such as carboxylic acids, hydroxyls and amines. Surface derivatization of the modified polymer film is achieved through the specific attachment of bioactive, immunoactive, electroactive, and catalytic agents to the surface of the electroactive polymer. A polyacetylene, polypyrrole, polyanilane, or polythiophene is modified to contain functional groups. When an analyte in a sample reacts with an indicator reagent, the electrical conductivity of the polymer is changed. The presence of the analyte is indicated by the change in electrical conductivity.

5,489,400. Molecular complex of conductive polymer and polyelectrolyte and a process of producing same. Issued: Feb. 6, 1996.

Inventors: Jia Liu, Linfeng Sun, and Sze Yang. Assigned to the Industrial Technology Research Institute. A processable, electrically conductive polymer composition contains a molecular complex made by template-guided chemical polymerization. The material has a polyelectrolyte and a conductive polymer.

5,536,808. Thiazole polymers and method of producing same. Issued: July 16, 1996.

Inventors: David Curtis and John Nanos. Assigned to Regents of the University of Michigan, Ann Arbor, MI. The patent reviews polymers that contain various cyclic units. Each of the units is connected to another thiazole unit.

5,569,708. Self-doped polymers. Issued: Oct. 29, 1996.

Inventors: Fred Wudl and Alan Heeger. Assigned to Regents of the University of California, Oakland, CA. A self-doped conducting polymer contains along its backbone an electron conjugated system that comprises several monomer units.

5,571,454. Soluble and processable doped electrically conductive polymer and polymer blend thereof. Issued: Nov. 5, 1996.

Inventors: Show-An Chen and Mu-Yi Hua. Assigned to Science Council. This invention relates to a homogeneously doped conductive polymer and a polymer blend in which the dopant is a protonic acid with a long carbon chain or larger substituent group. The resulting doped conductive polymer solution and polymer blend solution can be cast to form free-standing films with metallic luster, of which the conductivities remain stable in ambient conditions. The conductive polymer film and the conductive polymer composite film can be utilized in antistatic wrapping materials for electronic components and in EMI shielding.

5,589,108. Soluble alkoxy group–substituted aminobenzenesulfonic acid aniline conducting polymers. Issued: Dec. 31, 1996.

Inventors: Shigeru Shimizu et al. Assigned to Nitto Chemical Industry Co. Ltd., Tokyo, Japan. A soluble aniline conductive polymer comprises as a repeating unit an alkoxyl group–substituted aminobenzenesulfonic acid, its alkali metal salts, ammonium salts, and/or substituted ammonium salts. The material is a solid with an average molecular weight of about 1900 or more at room temperature.

5,589,565. Water-soluble conducting polyphenylene vinylene polymers. Issued: Dec. 31, 1996.

Inventors: Fred Wudl and Alan Heeger. Assigned to Regents of the University of California, Oakland, CA. A self-doped conductive polymer has along its backbone an electron-conjugated system. A conductive zwitterionic polymer is also provided, as are monomers used for preparing the polymer and electrodes comprising the polymer.

5,591,318. Method of fabricating a conductive polymer energy storage device. Issued: Jan. 7, 1997.

Inventors: Changming Li, Ke Lian, and Han Wu. Assigned to Motorola Energy Systems Inc., Northbrook, IL. A method for making high-power electrochemical charge storage devices. The process entails depositing an electrically conductive polymer onto a non-noble metal substrate that has been prepared by treatment with a surfactant. Using this method, high-power, high-energy electrochemical charge storage devices are fabricated at low cost.

5,595,689. Highly conductive polymer blends with intrinsically conductive polymers. Issued: Jan. 21, 1997.

Inventors: Vaman Kulkarni and John Campbell. Assigned to Americhem Inc., Cuyahoga Falls, OH. A polymer blend includes an intrinsically conductive polymer disbursed in a matrix selected from thermoplastic polymers, monomers, polymerizable precursors, and combinations thereof. The blend is improved by an amount of a nonpolymeric polar additive. The blend includes a matrix material selected from thermoplastic polymers, monomers, and polymer precursors and an intrinsically conductive polymer and a nonpolymeric highly polar additive, disbursed into a polymer blend.

5,633,098. Batteries containing single-ion conducting solid polymer electrolytes. Issued: May 27, 1997.

Inventors: Subhash Narang and Susana Ventura. Assigned to SRI International, Menlo Park, CA. Novel batteries containing single-ion conductive polymer electrolytes. The polymers are polysiloxanes substituted with fluorinated poly(alkylene oxide) side chains having associated ionic species.

5,637,421. Completely polymeric charge-storage device and method for producing same. Issued June 10, 1997.

Inventors: Theodore Poehler, et al. Assigned to Johns Hopkins University, Baltimore, MD. This patent discusses a technique for producing a quasi-solid state charge storage device capable of being repeatedly charged and discharged. The device has one or more electrochemical cells. Each cell is composed entirely of an ionically conductive gel polymer electrolyte layer. The layer separates opposing surfaces of electronically conductive conjugated polymeric anode and cathode electrodes that are supported on lightweight porous substrates.

5,645,890. Prevention of corrosion with polyaniline. Issued: July 8, 1997.

Inventors: Alan MacDiarmid and Naseer Ahmad. Assigned to Trustees of the University of Pennsylvania, Philadelphia, PA. This patent reviews methods for improving the corrosion inhibition of a metal or metal alloy substrate surface. The substrate surface is coated with a polyaniline film. The polyaniline film coating is applied by contacting the substrate surface with a solution of polyaniline. The polyaniline is dissolved in an appropriate organic solvent, and the solvent is allowed to evaporate from the substrate surface yielding the polyaniline film coating.

5,674,752. Conductive polymer coated fabrics for chemical sensing. Issued: Oct. 7, 1997.

Inventors: Leonard Buckley and Greg Collins. Assigned to The United States of America as represented by the Secretary of the Navy, Washington, DC. A fabric chemical sensor, a process, and an apparatus are disclosed. The sensor, process, and apparatus are for the detection, classification, identification, or quantification of one or more component chemicals of a chemical vapor. The determination is performed by a resistance measurement made across the sensor in response to exposure of the sensor to the chemical vapor.

5,681,442. Method of producing an organic device. Issued: Oct. 28, 1997.

Inventors: Kazufumi Ogawa and Norihisa Mino. Assigned to Matsushita Electric Industrial Co., Ltd., Osaka, Japan. A process for producing an organic device that entails forming a first electrode and a second electrode on a substrate with an insulating film formed on its surface, then etching out the insulating film with the first electrode and the second electrode used as masks. The process also involves spreading a surfactant containing the electrolytically polymerizable unsaturated groups on the surface of a water bath to form a monomolecular film.

Copies of patents may be obtained from Commissioner of Patents and Trademarks, Washington, DC 20231. Include number and $3.00 for each patent. ■

APPENDIX B. RESOURCES AND BIBLIOGRAPHY

Akagi, K., et al. 1995. Synthesis and properties of liquid crystalline polyacetylene derivatives. *Synthetic Metals,* 69, 13–16.

Dylis, D., et al. November 1996. All-plastic battery system for small electronic device applications. NASA Technology 2006 Conference Proceedings.

Epstein, A. J. June 1997. Electrically conducting polymers: science and technology. *MRS Bulletin,* 16–23.

Fahlman, M., et al. April 1997. Corrosion protection of iron/steel by polyanilines: a photoelectron spectroscopy study. Proceedings of the Society of Plastics Engineers, Annual Technical Conference, Toronto, Ontario, Canada.

Josowicz, M. April 1995. Applications of conducting polymers in potentiometric sensors. *The Analyst,* 120, 1019–1024.

Kathirgamanathan, P. and B. Boland. 1993. Direct electrodeposition of metals and conducting polymers on nonwoven thermoplastics on a continuous basis. *Journal of the Electrochemical Society,* 140(10), 2815–2818.

Killian, J., et al. 1996. Polypyrrole composite electrodes in an all-polymer battery system. *Journal of the Electrochemical Society,* 143(3), 936–942.

Linfeng, S., et al. 1997. Double-strand polyaniline. *Synthetic Metals,* 84, 67–68.

Musfeldt, J. L., et al. 1994. Luminescent polymers with discrete emitter units. *Journal of Polymer Science: Part B: Polymer Physics,* 32, 2395–2404.

Schoch, K. F. and H. E. Saunders. June 1992. Conducting polymers. *IEEE Spectrum,* 52–55.

Tolbert, L. 1992. Solitons in a box: the organic chemistry of electrically conducting polyenes. *Accounts of Chemical Research,* 25, 561-568.

Wang, Y. Z., et al. 1996. Alternating-current light-emitting devices based on conjugated polymers. *Applied Physics Letters,* 68(7), 894–896.

CLIENT STUDIES

TWENTY-FIVE EMERGING TECHNOLOGIES: DRIVING NEW MARKETS THROUGH THE 1990S

Selected from more than 200 strategic emerging technologies identified by Technical Insights' analysts, this report studies those with the greatest commercial potential for significant market development. Each promises substantial impact either on a major industry or industry segment. Every briefing gives: technical description, current status, potential impact, barriers to commercialization, likely time frame, market forecast. Report offers access to new markets via licenses, joint ventures, acquisition, by including full contact information of developers. Topics are:*

Rapid Metal Tooling

Recombinant Protein Manufacture in Plants

Miniature Imaging Devices

Electrostrictive Polymers

Field Emission Displays (FEDs)

Glass Material Oxidation and Dissolution (GMODS)

Chemical Mechanical Polishing (CMP)

Holographic Structured Light Generator

Scanning Probe Microscopy (SPM)

Alkali Metal Thermoelectric Converters (AMTEC)

Combinatorial Materials Discovery

Metal Contaminant Remediation

Luminescent Silicon

Engineered Response Gels (ERGs)

Precision Micromachining

Solid-Phase Epitaxial GaAs

Superconductive Electric Power Management

Radiance Process Surface Cleaning

Electrochemical Process Treats Hazardous Waste

Planar Optic Displays

Active-Pixel Sensors

Polymer Brushes

Biocatalytic Desulfurization (BDS)

Organically Modified Silicates

Polymer Multilayer Processes

*Fall 1997 . . . .$2850 (*Also sold separately for $450 each)*

Manufacturing On The Internet: Use 21st Century Techniques To Speed Your Product Cycles

The Internet has dramatically influenced manufacturing, bringing new benefits to a wide range of industries. This, 144 page, study puts you on the frontline of Internet developments. It deals with topics such as virtual prototyping and manufacturing, remote control of machinery, concurrent engineering and collaboration, CAD/CAM and product data management systems, procurement, conferencing, setting up intranets, researching manufacturing sites, and other ways to gain a competitive edge. If you're exploring ways to use the Internet, this report will show you how to make the most of your Internet investment.

Winter 1998 . . . .*$995*

Nanophase Materials: The Coming Boom In Catalysts, Cosmetics, Coatings, And Drugs

Nanophase materials take on new and unexpected properties arising from their small size and high surface area. For manufacturers, smaller materials can make a big difference in their products. New applications, new opportunities are ready to be exploited. The new materials are likely to have a significant impact in films and coatings, data storage, information displays, semiconductors, polymers, and other areas. This, 117 page, study is an invaluable guide to the small world of nanophase materials and structures.

Fall 1997 . . . .*$2450*

Frontiers In Food Science

This report uncovers the most important new research on food and nutrition. It taps the top authorities writing in the leading sources of medical and nutrition information. The latest and best intelligence has been uncovered and extracted into concise briefings. Gets you up to date on the latest research discoveries and guidelines to prevent food-related illness and death. At the same time, this unique, 488 page, report reduces the amount of reading and searching you have to do.

Fall 1997 . . . .*$295*

Lab-On-A-Chip: The Revolution In Portable Instrumentation, 2nd Edition

This 141 page completely revised second edition tracks at least two dozen research groups worldwide that are involved in developing the laboratory on a chip. About 50 companies are involved in developing the components needed: microelectromechanical devices (MEMs), microsensors, biosensors, nonfabricated gears, valves, pumps, micro-machined channels, and sample holders. Each of these concepts is discussed in-depth. Steers the way on how the technology can be exploited.

Fall 1997 . . . .*$2200*

Silicon Germanium: Key Technology for Digital Communications Expansion

This exclusive report is the first comprehensive overview to separate the real from the hype of this emerging technology. It explains the new technology, the solutions and the challenges it offers to developers of high-speed electronic devices. Silicon Germanium spotlights the technology programs and offerings of 22 companies, and the work of 34 universities and research institutions worldwide. It explores the prospects for Silicon Germanium (SiGe) in the marketplace, and the most promising opportunities in this exploding market. Provides names, addresses, phone and fax numbers and e-mail addresses of the principal players so you can contact them directly. Also provides a list of key patent holders.

Summer 1997 . . . .*$2250*

EXTREMOZYMES AND COMMERCIALLY IMPORTANT EXTREMOPHILES: THE NEXT WAVE IN INDUSTRIAL MANUFACTURING

A comprehensive report on the new frontier in biotechnology: enzymes and microorganisms that thrive in hostile environments like hot springs, arctic regions and intense u.v., and promise to revolutionize applications ranging from laundry detergents to oil recovery, from food to drug manufacture. Over twenty groups are working on techniques to use them, and an explosion on "green" manufacturing is in the making.

Summer 1997 . . . .*$1995*

NEURAL NETS IN MANUFACTURING: SMART COMPUTING FOR REAL WORLD PROBLEMS

Neural Nets – control systems that "learn" and can closely emulate human decision-making processes – are fast becoming an industrial reality. This report covers the work of 30 research groups who are applying neural nets to operations as diverse as factory floor process control to ore extraction. Quality assurance, motion control, engine diagnostics, casting, molding, will benefit from this smart approach to applied intelligence.

Summer 1997 . . . .*$1800*

CORPORATE GROWTH STRATEGIES: HOW TO FIND/EXPLOIT NEW TECHNOLOGY AND PRODUCTS, 4TH EDITION

Comprehensive guide to strategies and sources of continuous stream of outside technology necessary for corporate growth. Analyzes all aspects of acquisition, partnering, technology transfer, coventuring, licensing. Provides contact names, addresses worldwide. Outlines: steps in staffing for R&D, identifying outside R&D, monitoring competition, how to take equity stakes, acquisition strategies, understanding mindset of academic researchers, how to profit from government research, caveats of licensing agreements, international opportunities, discusses venture capital possibilities, and much more.

Spring 1997 . . . .*$1620*

FAST CHIPS/SUPERSTORAGE: OPPORTUNITIES FROM THE NEXT GENERATION OF INFORMATION TECHNOLOGIES

This report examines changes expected in the world of information-processing hardware in the next ten years. Likely advances in chip-making materials and processes, new data storage technologies and new classes of smart sensing systems are previewed. With these advances will come manufacturing, marketing, and investment opportunities through licenses, partnerships, acquisitions, distribution agreements and other arrangements. This single-volume resource will quickly prepare you to profit from near-term and mid-term developments in these emerging fields. Provides the names, addresses and phone numbers of the individual researchers under discussion.

Spring 1997 . . . .*$2125*

GROWTH OPPORTUNITIES IN BIOMEDICAL DEVICES: KEY DEVELOPMENTS WITH NEAR-TERM COMMERCIAL POTENTIAL

This report is a far-ranging guide to current and future opportunities in biomedical devices. It takes you inside corporate, academic and government laboratories around the world to give you first hand intelligence on emerging technologies with true commercial potential. It briefs you on hundreds of advances from diagnostics to monitoring systems to tele-medicine and many more. Assesses the likely market impact, competitive technologies and barriers to development for each technology covered and provides names, addresses, and phone numbers of the laboratory heads and individual researchers involved in each development.

Winter 1997 . . . .*$2250*

EASY FABRICATION OF HIGH-TECH MATERIALS: A NEW ERA OF WIDESPREAD COMMERCIALIZATION

This report is a comprehensive guide to the new era in high-tech materials. The cost and usability of many high-tech materials are changing thanks to rapid advances in dozens of fabrication technologies, positioning high-tech metals, ceramics, polymers and other materials to be used in a wide range of commercial, consumer and industrial applications. This report gives in-depth technical explanations suited to managers of R&D, finance, marketing and other disciplines. It provides real-world views of commercial potential and includes contact names, addresses, and phone numbers.

Fall 1996 . . . .*$1785*

THROWAWAY SENSORS: ULTRA-LOW COST SENSORS KEY TO SMART PRODUCTS, SMARTER MANUFACTURING

This exhaustive report is a guide to ultra-low cost sensors, sensors built using micromachining and advanced techniques that provide full sensing capabilities at a fraction of the cost of conventional devices. It brings together hundreds of separate bits of intelligence to give both an overview and a level of detail unavailable from any other source. This report gives you a complete business perspective on the enormous commercial impact of this emerging field; provides a guide to the key players including names, addresses and phone numbers of principal individuals; major patents; and directories of related resources.

Fall 1996 . . . .*$1775*

AEROGELS AND XEROGELS: GROWTH AND OPPORTUNITIES FOR THE EARLY 21ST CENTURY

This report examines the rapidly expanding world of aerogels and their chemical cousins, xerogels. With astoundingly useful properties and a growing list of commercial applications, these advanced materials are poised to have a major impact. This report takes you into every major center working on aerogels (from startup companies to household-name organizations) and presents its information in a form designed specifically to guide you through the decision-making process. Includes the names, mail and e-mail addresses, phone and fax numbers of all the researchers it reports on, as well as key patents and an exhaustive bibliography.

Summer 1996 . . . .*$1925*

Environmental Sensors and Monitoring: Technologies and Opportunities, 2nd Edition

A greatly expanded step-by-step guide through a complex group of biosensor, optic sensor and mass sensor technologies, patents and regulations. It analyzes the forces driving the markets for underground storage tank monitoring, water, air, and land quality. Organizations and companies involved in the most promising research highlighted. Names, telephone and fax numbers and even World Wide Web addresses are given, so you can make immediate contact, on your own, to get the information you need.

Spring 1996 . . . . *$1150*

Managing Innovation for Profit, 7th edition

A compilation of more than 100 in-depth analyses of all aspects of R&D. This strategic management tool contains tested techniques, specific guidelines, proven strategies, and hands-on instruction from leading experts at major consulting firms, manufacturing companies, universities, government agencies. Learn how to confront successfully such critical issues as product development, R&D productivity, competitive intelligence, R&D planning, and measurement. This report shows you how to find and manage the new technology that is the key to your company's future survival and growth. Looseleaf-bound and tabulated for easy reference.

Spring 1996 . . . . *$1100*

Light-Emitting Polymers: The Technology and Opportunities

Light-Emitting Polymers (LEPs) are conjugated polymers, a class of electrically conductive materials that are flexible and wafer-thin – and emit the full spectrum of visible light when a current is passed through them. This comprehensive report provides a complete update on the science and business of LEPs plus a framework for profiting from near-term manufacturing, marketing, licensing, and investment opportunities.

Spring 1996 . . . . *$1625*

Electronic Intelligence

TECHNICAL INSIGHTS ALERT® is a weekly information service that tells you about new and significant advances in technology and how you can get access to them. The information is delivered to you electronically via e-mail, Lotus Notes or HTML format available for distribution at your site or throughout the company.

For 12 critical areas — industrial R&D, high-tech materials, coatings technology, advanced manufacturing, sensor technology, industrial bioprocessing, genetic technology, microelectronics technology, chemical process, bio/med technology, food technology and nanotechnology — ALERT provides a screened, weekly flow of concise briefings of 250–350 words each. They are intended for quick scanning by R&D managers, technical managers, new product teams, technology scouts, project leaders, etc.

Every ALERT briefing is complete with . . . explanation of a new technology . . . potential market impact . . . technology transfer arrangements sought by the developers . . . important patents issued . . . complete contact details, including: names, mail and e-mail addresses, phone and fax numbers for your follow-up.

A thirteenth area is covered — emerging technologies. This service provides an in-depth analysis of a business opportunity. This 14 to 24 page briefing identifies one critical technology a month.

TECHNICAL INSIGHTS ALERT is designed to take optimal advantage of your company's internal communication facilities such as electronic mail networks. With a corporate license, you can route vital technology intelligence reports instantly to everyone who can benefit from them throughout your organization.

INTELLIGENCE SERVICES:

Advanced Coatings & Surface Technology	12	$620	$680
Advanced Manufacturing Technology	12	$660	$720
Bio/Med Technology Alert	52	$890	$990
Chemical Process Alert	52	$890	$990
Food Safety Notebook	12	$95	$115
Futuretech	12	$1500	$1600
Genetic Technology News	52	$845	$945
High-Tech Materials Alert®	12	$675	$735
Industrial Bioprocessing	12	$625	$685
Inside R&D®	52	$840	$940
Microelectronics Technology Alert	52	$890	$990
Nutrition Research Newsletter	12	$149	$169
Science & Government Report	20	$490	$570
Sensor Technology	12	$650	$710

FOR MORE INFORMATION, CONTACT LYN SCHMIDT AT 201-568-4744, EXT. 238;
FAX: 201-568-8247; INTERNET: LYNRRE@INSIGHTS.COM; COMPUSERVE: 73373,51
URL: HTTP://WWW.INSIGHTS.COM

TECHNICAL
INSIGHTS
John Wiley & Sons, Inc.

⊛WILEY

Publishers Since 1807

Research services for senior management and technical executives, worldwide, concerned with planning R&D, new ventures, new product development, or with corporate strategy.

INTELLIGENCE SERVICES

Inside R&D®

Genetic Technology News

Advanced Manufacturing Technology

Sensor Technology

High-Tech Materials Alert®

Advanced Coatings & Surface Technology

Industrial Bioprocessing

Chemical Process Alert

Science & Government Report

Microelectronics Technology Alert

Bio/Med Technology Alert

Food Safety Notebook

Nutrition Research Newsletter

IN-DEPTH REPORTS — EMERGING TECHNOLOGIES AND CLIENT STUDIES

Future-oriented assessments of new, significant technologies — their potentials, their threats.

FUTURETECH

A service tracking strategic technologies.

ELECTRONIC SERVICES

TECHNICAL INSIGHTS ALERT© — an electronic information service reporting on significant technological advances and the industrial markets they will influence.

Technical Insights
John Wiley & Sons, Inc.
32 North Dean Street
Englewood, NJ 07631-2807

Phone: 201-568-4744 • Fax: 201-568-8247
E-Mail: reports@insights.com
URL: http://www.insights.com